SpringerBriefs in Molecular Science

# Chemistry of Foods

**Series Editors**

Salvatore Parisi, Lourdes Matha Institute of Hotel Management and Catering Technology, Thiruvananthapuram, Kerala, India

Ricardo Pereira, Centre of Biological Engineering, University of Minho, Braga, Portugal

The series Springer Briefs in Molecular Science: Chemistry of Foods presents compact topical volumes in the area of food chemistry. The series has a clear focus on the chemistry and chemical aspects of foods, topics such as the physics or biology of foods are not part of its scope. The Briefs volumes in the series aim at presenting chemical background information or an introduction and clear-cut overview on the chemistry related to specific topics in this area. Typical topics thus include:

- Compound classes in foods—their chemistry and properties with respect to the foods (e.g. sugars, proteins, fats, minerals, …)
- Contaminants and additives in foods—their chemistry and chemical transformations
- Chemical analysis and monitoring of foods
- Chemical transformations in foods, evolution and alterations of chemicals in foods, interactions between food and its packaging materials, chemical aspects of the food production processes
- Chemistry and the food industry—from safety protocols to modern food production

The treated subjects will particularly appeal to professionals and researchers concerned with food chemistry. Many volume topics address professionals and current problems in the food industry, but will also be interesting for readers generally concerned with the chemistry of foods. With the unique format and character of SpringerBriefs (50 to 125 pages), the volumes are compact and easily digestible. Briefs allow authors to present their ideas and readers to absorb them with minimal time investment. Briefs will be published as part of Springer's eBook collection, with millions of users worldwide. In addition, Briefs will be available for individual print and electronic purchase. Briefs are characterized by fast, global electronic dissemination, standard publishing contracts, easy-to-use manuscript preparation and formatting guidelines, and expedited production schedules.

Both solicited and unsolicited manuscripts focusing on food chemistry are considered for publication in this series. Submitted manuscripts will be reviewed and decided by the series editor, Prof. Dr. Salvatore Parisi.

To submit a proposal or request further information, please contact Dr. Sofia Costa, Publishing Editor, via sofia.costa@springer.com or Prof. Dr. Salvatore Parisi, Book Series Editor, via drparisi@inwind.it or drsalparisi5@gmail.com

Salvatore Parisi

# Nutrition, Chemistry, and Health Effects of Sugar, Salt, and Milkfat

 Springer

Salvatore Parisi
Lourdes Matha Institute of Hotel
Management and Catering Technology
Thiruvananthapuram, Kerala, India

ISSN 2191-5407          ISSN 2191-5415  (electronic)
SpringerBriefs in Molecular Science
ISSN 2199-689X          ISSN 2199-7209  (electronic)
Chemistry of Foods
ISBN 978-3-031-67394-8          ISBN 978-3-031-67395-5  (eBook)
https://doi.org/10.1007/978-3-031-67395-5

This Springer imprint is published by the registered company Springer Nature Switzerland AG
The registered company address is: Gewerbestrasse 11, 6330 Cham, Switzerland

If disposing of this product, please recycle the paper.

# Acknowledgments

This book is the 45<sup>th</sup> title of the *Springer Briefs in Molecular Science: Chemistry of Foods* series: it can be published exactly 10 years after the first book in 2014, demonstrating that the scientific interest in food chemistry and related disciplines continues to remain notable. I wish to mention and thank Dr. Ramesh Kumar Sharma (Food Safety Consultant and Scientific Writer, Tilam Sangh Rajasthan, Bikaner, Rajasthan, India) who conceived the original project of the book (May 2022) and its initial development.

Palermo, Italy
May 2024

Salvatore Parisi
Series Editor,
*Springer Briefs in Molecular Science: Chemistry of Foods*

# Contents

# Chapter 1
# Sugar, Salt, Milkfat as Flavour and Satiety Ingredients. A Chemical Perspective

**Abstract** The modern industry of prepared foods has to face some important challenges when speaking of taste, palatability, and satiety. From the practical viewpoint of food producers, the following factors should be considered: the approximate composition of products; target of food consumers; considered marketing channels; type of post-production services; and so on. In this specific ambit, desired sensorial properties should be obtained by means of the modification of the original recipe or idea of the food product, also considering enhanced features such as long durability and ameliorated colours. In general, the research for tasteful, sweet, palatable foods (and also for products able to promote satiety) takes into account the role of three basic ingredients or classes of ingredients for food production: sugars, salt, and milkfat. Different perspectives can be considered when speaking of these ingredients and their influence of sensorial properties of foods and beverages. This chapter is dedicated to a general description of chemical features of these compounds and classes of ingredients and related effects when speaking of edible preparations. The chemical characterisation can notably enhance the comprehension of related uses in the food and beverage sector. The following categories—sugars, including several similar definitions for these molecules; food-grade salt/sodium chloride, with different types including fortified versions for nutritional and health purposes; milkfat—are discussed with relation to chemical and physical features, and specific characteristics linked to palatability and satiety performances, also taking into account current Codex Standards.

**Keywords** Added sugar · Disaccharide · Monosaccharide · Palatability · Satiety · Sodium chloride · Triacylglycerol

## Abbreviations

| | |
|---|---|
| DSD | Disaccharide |
| EUFIC | European Food Information Council |
| FAO | Food and Agriculture Organization of the United Nations |

| | |
|---|---|
| F&B | Food and beverage |
| FDA | Food and Drug Administration |
| HACCP | Hazard Analysis and Critical Control Points |
| IFST | Institute of Food Science & Technology |
| IFT | Institute of Food Technologists |
| IDFA | International Dairy Foods Association |
| MSD | Monosaccharide |
| NaCl | Sodium chloride |
| WISDOM | Water Incorporation, Shelf–life Deduction and Orientation in the Modeling |
| WHO | World Health Organization. |

## 1.1  Introduction

The modern industry of prepared foods has to face some important challenges when speaking of taste, palatability, and satiety. The basic problem of food and also beverage products, from the specific viewpoint of marketing strategists, is to match current desires of food consumers, and this strategy should be (Ackroff and Sclafani 1999; Drewnowski and Monsivais 2020; Gates et al. 2007; Ha et al. 2018; Just et al. 2014; Klosse et al. 2004; Lambert et al. 2024; Young 2021):

(a) Applicable to the normal food production with increase of produced items (in terms of amounts of sold units) and decrease of production costs.
(b) Easily 'translatable' in a practical and comprehensible marketing/advertising approach.
(c) And, last but not least, ameliorable from time to time, taking into account possible fluctuations of prices (raw materials, intermediates, packaging materials, delivery services, etc.) and new technologies/products such as cultivated meats.

From the practical viewpoint of food producers, the above-mentioned strategy should take into account the approximate composition of products, also on the basis of several factors: target of food consumers; considered marketing channels; type of post-production services; and so on. In this specific ambit, desired sensorial properties should be obtained by means of the modification of the original recipe or idea of the food product, also considering enhanced features such as long durability and ameliorated colours (Delahunty and Havermans 2010; Lillford 2016; Low et al. 2014; Yeomans 1998).

In general, the research for tasteful, sweet, palatable foods (and also for products able to promote satiety) takes into account the role of three basic ingredients or classes of ingredients for food production: sugars, salt, and milkfat (Belc et al. 2019; Gibson and Ashwell 2011; Mahato et al. 2020; Mukhopadhyay et al. 2020).

The first category—sugars—is variegated enough, and it is substantially a subgroup of carbohydrates: as a result, a detailed and separated discussion has to be needed. The same thing is true with reference to milkfat (onwards named milkfat), meaning that a notable number of fat and fat-related molecules are included in this broad category.

On the other side, food-grade salt is only one molecular and ionic compound: sodium chloride. However, different commercial types of food-grade salt are available because of their rock or sea origin, geographical characterisation, different sensorial attributions (colour, taste, etc.), possible working processed on salt (as an example, smoked salt specialties), and other features which could be covered by adequate certification standards. Consequently, 'salt' should be characterised in detail.

Different perspectives can be considered when speaking of these ingredients and their influence of sensorial properties of foods and beverages. This chapter is dedicated to a general description of chemical features of these compounds and classes of ingredients and related effects when speaking of edible preparations. The chemical characterisation can notably enhance the comprehension of related uses in the food and beverage (F&B) sector.

## 1.2  Sugar in Food Products and Preparations. Basic Chemical Features

The list of 'sugars' for F&B purposes can be long enough. Basically, sugars—also named 'added sugars', 'sugar', and 'caloric sweeteners' in the United States of America at least (Sigman-Grant and Morita 2003)—are part of the 'carbohydrate' group of nutrients, but the distinction between different sugars cannot be defined without other definitions. First of all, 'sugars' are mono- or di-saccharides, part of the broad group of carbohydrates.

Chemically speaking, carbohydrates have the general molecular formula $C_x(H_2O)_y$, and these molecules have hydroxyl and carbonyl groups in their structure. As a result, these molecules can be defined complex aldehydes or ketones depending on the functional carbonyl group (Stylianopoulos 2013). Because of this peculiar group, linear structures of carbohydrates are also represented as ringer structures when in solution, provided that five, six, or sevencarbon atoms are present. Should this condition be verified, optical activity would be observed with a peculiar $\alpha$- or $\beta$-configuration depending on thermodynamic stability (Sigman-Grant and Morita 2003). Consequently, carbohydrates could be also considered as a succession of ringed structures called monomers or monosaccharides.

Monosaccharides are the simplest example of sugars (three to seven carbon atoms are involved, while the remaining atoms are oxygen and hydrogen). For this reason and the consequent low molecular weight value, monosaccharides (MSD) represent the absorbable source of sugars. Should two MSDs be linked together and consequently involved in the general carbohydrate structure, a disaccharide (DSD) structure would be considered. As a result, the group of MSD and DSD is generally considered as 'sugars' or 'simple sugars'. However, some additional description is needed.

In fact, should a MSD number between three and nine be involved in the structure of carbohydrates, an oligosaccharide would be obtained. Consequently, the group of complex carbohydrates with a name ending as 'saccharides' is at least a tripartite category. In general, the description of similar molecules for food-grade uses should be explained as follows (Clemens et al. 2016; Coultate 2009; Davis 1995; Scapin et al. 2017; Schorin et al. 2012; Sigman-Grant and Morita 2003; Wrolstad 2012):

(1) 'Sugars' or 'simple sugars' are MSD a DSD only.
(2) Alternatively, a 'sugar' definition (no plural meanings) should be ascribed to sucrose only (sucrose, is one of the most important DSD representatives).
(3) Oligosaccharides refer to carbohydrates ranging from three to nine MSD rings.
(4) The 'added sugars' term refers to all possible MSD, DSD, and oligosaccharides intended as raw materials (mixtures of carbohydrates and other compounds). In this situation, the following raw materials are included: white, brown, and raw sugar; corn syrup, high-fructose corn syrup, other syrups including malt type; fructose (as liquid material and sweetener); dextrose (anhydrous and crystal types); honey; molasses; etc. Oligosaccharides are a minor portion if compared with MSD and DSD.
(5) 'Caloric sweeteners' refer to molecules able to give a sweetening effect to food and beverage products. Once more, these additives can be single compounds (sucrose, dextrose, etc.) or complex mixtures such as honey and syrups; consequently, oligosaccharides can be included.

Such a classification highlights the main role of sucrose, one of DSD compounds, because this molecule—extracted from sugarcane or beets—is extremely sweet; as a result, its concentration is considered as a conventional measure for sweetening potency (Kim et al. 2014; Starkey et al. 2022; Su 2000). Anyway, the basic MSD are (Figs. 1.1 and 1.2):

(a) β-*D*-glucose (also named aldohexose and dextrose) with dextrorotatory optical activity;

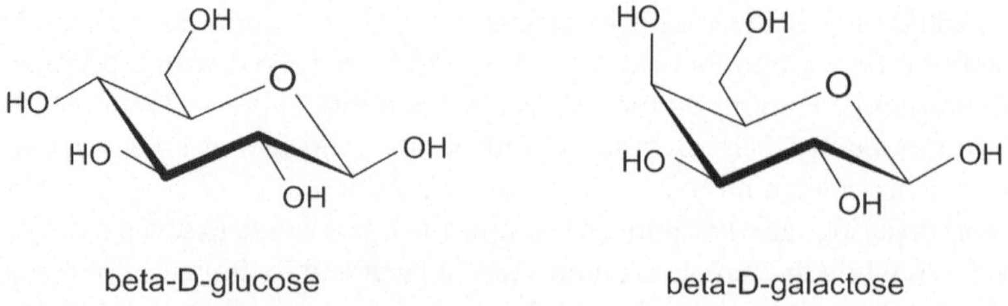

beta-D-glucose                    beta-D-galactose

**Fig. 1.1** Molecular structures of basic MSD: β-*D*-glucose and β-*D*-galactose. BKchem version 0.13.0, 2009 (http://bkchem.zirael.org/index.html) has been used for drawing this structure

beta-D-mannose                    beta-D-fructose

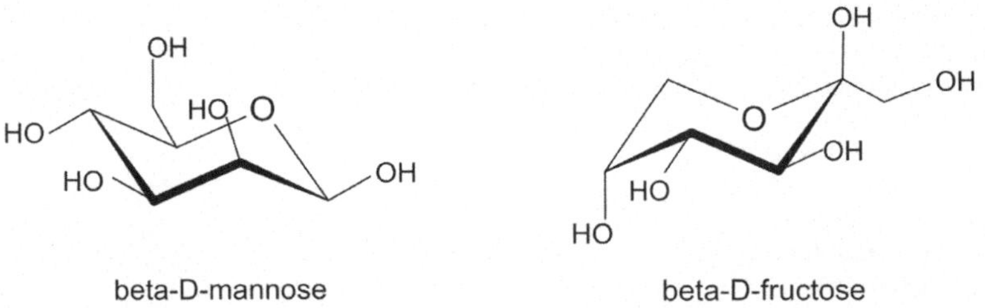

**Fig. 1.2** Molecular structures of basic MSD: β-*D*-mannose and β-*D*-fructose. BKchem version 0.13.0, 2009 (http://bkchem.zirael.org/index.html) has been used for drawing this structure

(b)  β-*D*-galactose;
(c)  β-*D*-mannose;
(d)  β-*D*-fructose (a ketohexose),

where the last compound is not a 6-carbon ring: the structure is a 5-carbon ring, being it a furanose with a functional ketone group on the second carbon atom of the linear chain (Sigman-Grant and Morita 2003). The first of carbon atoms on the linear chain, C1, is also named anomeric carbon; consequently, α- and β-isomers are also named 'anomers'.

With reference to DSD, four main molecules should be considered (Figs. 1.3, 1.4, 1.5 and 1.6):

(1)  Sucrose (obtained by joining one α-glucose ring and one β-fructose ring).
(2)  Lactose (obtained by joining one α-galactose ring and one α- or β-glucose ring).
(3)  Maltose (obtained by joining two α-glucose rings, with 1–4 linkage).
(4)  Trehalose (obtained by joining two α-glucose rings, with 1–1 linkage).

Oligosaccharides, containing three to nine MSD units, can give an idea of the qualitative difference between simple sugar mixtures and syrups. As an example, the minor portion of these oligosaccharides should not exceed 2.4% in fructose-abundant corn syrups (Sigman-Grant and Morita 2003). Anyway, some of these

**Fig. 1.3** Molecular structure of basic DSD: sucrose (obtained by joining one α-glucose ring and one β-fructose ring). BKchem version 0.13.0, 2009 (http://bkchem.zirael.org/index.html) has been used for drawing this structure

Sucrose

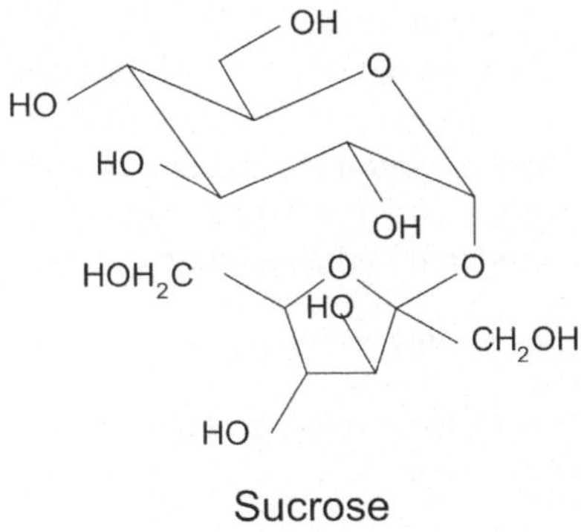

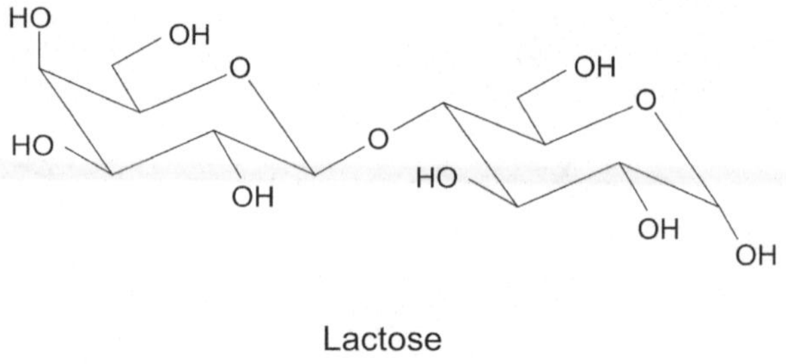

**Fig. 1.4** Molecular structure of basic DSD: lactose (obtained by joining one α-galactose ring and one α- or β-glucose ring). BKchem version 0.13.0, 2009 (http://bkchem.zirael.org/index.html) has been used for drawing this structure

Lactose

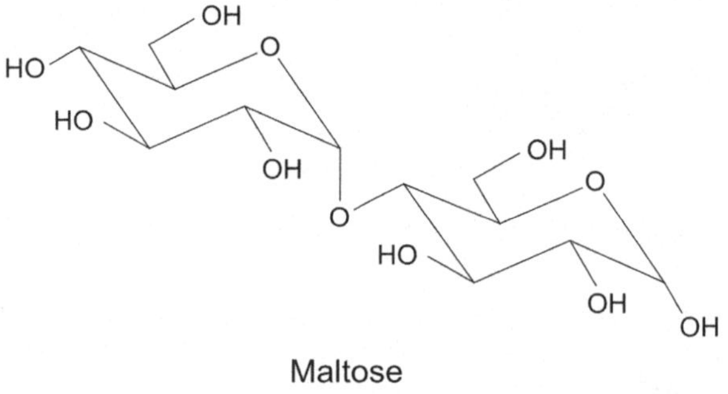

**Fig. 1.5** Molecular structure of basic DSD: maltose (obtained by joining two α-glucose rings, with 1–4 linkage). BKchem version 0.13.0, 2009 (http://bkchem.zirael.org/index.html) has been used for drawing this structure

Maltose

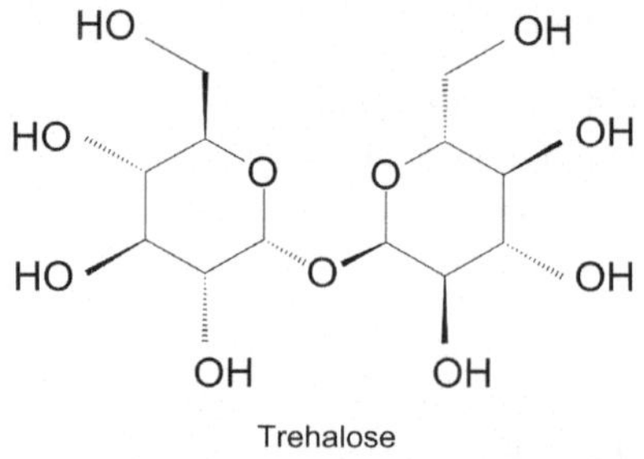

**Fig. 1.6** Molecular structure of basic DSD: trehalose (obtained by joining two α-glucose rings, with 1–1 linkage). BKchem version 0.13.0, 2009 (http://bkchem.zirael.org/index.html) has been used for drawing this structure

Trehalose

molecules are important enough when speaking of sweetness enhancement: lactosucrose, maltotriose, raffinose, and 4-galactosyl-kojibiose are reported in this ambit, but it has to be noted that these oligosaccharides have three MSD units only. Should MSD be > 3, the reported taste would generally be perceived as starchy instead of sweet. This digression is important enough because many sweet-enhancing compounds and additives are not 'sugars': in detail, the following species and molecules have to be considered in the ambit of sweetening potency (Starkey et al. 2022):

(a) Dietary fibre.
(b) Fructans.
(c) Galacto-oligosaccharides.
(d) Gums.

(e)  High-intensity sweeteners (include acesulfame potassium, advantame, cycla-
mate, saccharin, etc.).
(f)  Maltodextrins.
(g)  Resistant and digestible starches.
(h)  Sugar alcohols (sorbitol, xylitol, lactitol, mannitol, maltitol, sorbitol, etc.).
(i)  Sweetening dipeptides (aspartame, neotame, etc.).
(j)  Sweetening proteins (brazzein, miraculin, neoculin, thaumatin, etc.).
(k)  Terpenoids (glycyrrhizin, steviosides, etc.).

The extreme variability of these compounds and their different chemical classification (carbohydrates, proteins, synthetic sweeteners, etc.) have to be taken
into account: from the viewpoint of the F&B consumer, sweetness is a distinctive feature of 'sugars', while many of the above-mentioned non-sugar molecules
have really remarkable sweetening power. As an example, the following non-sugar
compounds—neotame and thaumatin—are 7000 and 2000 sweeter than sucrose
(sweetness power = 100), respectively (Starkey et al. 2022). However, this work
is specifically dedicated to the use of common sugars (MSD, DSD), and syrups.
Consequently, non-sugar compounds and additives are not further discussed.

Apart from this chemical classification, the following chemical and physical properties should be mentioned because of important consequences in the F&B sector for
some MSD and DSD. It has to be taken into account that some of these properties are
extended to a whole class of sugars instead of a single compound such as fructose and
trehalose (Feofilova et al. 2014; Hoffmann 2010; Parisi and Luo 2018; Parisi et al.
2019a; Shakoor et al. 2022; Singla et al. 2018; Tappy and Lê 2010; Twarda-Clapa
et al. 2022):

(1)  Reducing sugars in general (all MSD, some DSD, and oligosaccharides; on
the other hand, non-reducing DSDs have both anomeric carbons linked in the
glycosidic bond, and different oligosaccharides are non-reducing sugars): their
main reaction under heating processes is to react with protein and the consequent
formation of Maillard products via Schiff's bases production.
(2)  Invert sugar (a mixture of glucose and fructose from sucrose): this additive is
able to inhibit crystallisation.
(3)  Fructose:

Temperature, pH, the amount of solid matter, and also the concomitant presence of non-fructose sweeteners, are able to influence positively sweetening
power of fructose.
Fructose can absorb water with more efficacy than sucrose if relative humidity
values are low (consequently, fructose can be a good humectant agent).
In addition, being fructose able to absorb water, it can act as an antimicrobial
compound, also lowering freezing point of related mixtures.
Fructose can also be an excellent gel-forming agent if added to starch (low
gelatinisation temperature is required).

(4)  Trehalose: this sugar exhibits non-reducing features; in addition, it is not able
to react with protein and is highly resistant to hydrolytic reactions.

On these bases, it can be assumed that MSD, DSD, and also some oligosaccharides have interesting features when speaking of food technology. This reflection should be taken into account if nutritional evaluation and risk assessment on sugars are discussed (Chaps. 3, 4, and 5). In other terms, are sugars really needed by the viewpoint of F&B technologists and producers? Certainly, basic F&B features such as leavening promotion and consequent texture (caused by fermentation processes), rheological performances, effects linked with osmotic pressure, and caramelisation are normally expected when speaking of sugars (Davis 1995).

In addition, the technological origin of sugars may be useful when speaking of basic raw materials: sugar cane, corn, sugar beet, honey; etc. (Baglio 2014, 2017; Barbieri et al. 2014). Some preference exists when speaking of corn syrups (bakery foods, dairy products, candies, processed foods, and beverages), while beet or cane sugar is reported to have a broader use (canned, frozen, or bottled foods; other preparations, etc.) (Sigman-Grant and Morita 2003). The problem may be also linked to a commercial dispute: as reported in 2012, the Corn Refiners Association recently petitioned the Food and Drug Administration (FDA) to allow the replacement of high-fructose corn syrup with the more familiar 'corn sugar' name. The FDA rejected this bid because food consumers could be confused in this way. This situation demonstrates that the origin and consequent labelling of sugars have its own commercial importance, with reference to high-fructose corn syrup and other commonly used sweetening sugars (with technologically correct and 'complicated' names). Basically, the problem concerns public perception of F&B products, including qualifying ingredients or additives (Allstun 2013; Fulton and Charles 2012; Hsu 2012). For these reasons, the sector of 'sugars' may be confusing enough for non-experienced F&B consumers, and more information and training actions would be really needed in this ambit (Khandpur et al. 2020; Kyle and Thomas 2014; Patterson et al. 2012).

From the technical viewpoint, the interested Reader can consider the Codex Standard for Sugars (CXS 212–1999, lastly amended in 2022). This reference document (Codex Alimentarius Commission 1999a) concerns all sugar products for general purposes and also sold as foodstuff, without further processing. The list of defined commodities is long enough:

(1)  White sugar.
(2)  Plantation or mill white sugar.
(3)  Powdered sugar (icing sugar).
(4)  Soft white sugar.
(5)  Soft brown sugar.
(6)  Dextrose anhydrous.
(7)  Dextrose monohydrate.
(8)  Powdered dextrose (icing dextrose).
(9)  Glucose syrup.
(10)  Dried glucose syrup.
(11)  Lactose.
(12)  Fructose (laevulose).
(13)  Raw cane sugar.

Interestingly, all of these definitions contain general chemical features (polarisation value, basic composition, possible presence of anticaking agents, etc.). In addition, these commodities do not concern explicitly certain DSD, while lactose and sucrose have been cited. The document also gives additional composition and quality data concerning these products when applicable concerning loss on drying, colorimetric appearance, pH, starch content, sulphated ash, conductivity ash, and amounts of invert sugar and 'sucrose plus invert sugar'.

## 1.3 Salt in Food Products and Preparations

Food-grade salt is one specific ionic compound: sodium chloride. As previously discussed, different commercial types of food-grade salt are available. The basic discrimination can be discussed on the following bases:

(a) Technological origin: rock, sea, etc.
(b) Geographical characterisation.
(c) Different sensorial attributions (colour, taste, etc.).
(d) Processing steps (smoked salt specialties, etc.).
(e) Other features which could be assessed by adequate certification standards.

As a result, 'salt' can be produced and sold on the general market, and also on the market of food ingredients, under many forms, names, and so on. A detailed and chemical description would be needed, also because of the technological influence (and aims) in the F&B sector.

From the technical viewpoint, the interested Reader can consider the Codex Standard for Food Grade Salt (CODEX STAN 150–1995). This reference document (Codex Alimentarius Commission 1995a) concerns all food-grade salts for general purposes and also sold as foodstuff. Basically, the product corresponds to one single definition, but—as above mentioned—this definition is explicitly linked to the following origin sources:

(a) Waters—seas, lakes, and springs (Bozkaya and Aluç 2024).
(b) Underground rock salt deposits.
(c) Natural brine.

Additional origins would be excluded. In addition, the food-grade salt has to have a minimum amount of sodium chloride (NaCl) of 97% on dry matter, while additional components have to be 'natural secondary products' whose origin and reason for presence are linked to the origin source and the production methods. Basically, these products are:

(1) Sulphates of calcium, potassium, magnesium, and sodium.
(2) Carbonates of calcium, potassium, magnesium, and sodium.
(3) Bromides of calcium, potassium, magnesium, and sodium.
(4) Chlorides of calcium, potassium, and magnesium.

With reference to additional compounds, copper is explicitly mentioned: the related allowed amount cannot exceed 2 mg per kilogram of salt product. In addition, the Standard clearly defines the boundaries for different versions of food-grade salt (salt as carrier of different and nutritionally interesting nutrients) (Codex Alimentarius Commission 1999a, b; Greenwald et al. 2022):

(a) Curing salt (addition of nitrate and/or nitrite compounds).
(b) Fortified salts (addition of iron, vitamins, fluoride, iodate or iodide compounds, stabilising agents, etc.). In particular, iodine addition (iodine-fortified food-grade salts) is required in certain geographical areas where iodine deficiency is known. Anyway, the fortification is carried out with the addition of sodium and potassium iodates or iodides. Iodine levels have no fixed values, according to this Standard (no minimum or maximum allowed levels).

Moreover, general contaminants should be taken under control according to provisions of the Codex General Standard for Contaminants and Toxins in Foods and Feeds-CODEX/STAN 193–1995 (Codex Alimentarius Commission 1995b). Copper, arsenic, mercury, dead, and cadmium are explicitly mentioned when speaking of analytical methods for determination.

The name of the product is important enough: in fact, the following definitions— 'Food Grade Salt', 'Cooking Salt', or 'Table Salt'—are allowed. Other qualifying terms are allowed on condition that:

(1) Salt contains ferrocyanide salts (added to brine during crystallisation); in this situation, the 'dendritic' adjective can be added.
(2) Fortified salt; in this situation, the fortifying compound has to be clearly evident (e.g. 'iodated', 'iodised', 'fortified with iron', etc.) commonly allowed descriptions).
(3) Peculiar origin source and/or production method; in this situation, such a mention has not to mislead or deceive the food consumer.

Another important pre-condition concerning special/fortified salts concern iodine loss. In other words, the predictable amount of supplied iodine in iodine-fortified salts could be lower when expected. Consequently, some precautions have to be put in place (Codex Alimentarius Commission 1999a, b):

(a) Adequate food-grade only bags, including the following plastics: high-density polyethylene, and/or laminated or non-laminated polypropylene, and/or low-density polyethylene.
(b) Adequate packing dimensions: the maximum allowed weight should not exceed 50 kg.
(c) The distribution should be optimised (in other words, temporal period between iodine fortification and salt consumption should be minimised).
(d) Correct storage, transportation, and sale conditions: absence of rain and high environmental humidity, absence of direct sunlight exposure; adequate ventilation.

(e) Finally, the food consumer has to be correctly informed with reference to adequate storage advices (labelling information concerning moisture, direct sunlight, and heat exposure).

Apart from these descriptions, the problem of salt impurities can be a notable concern. It has been reported that moisture absorption and cracking defects are related to the presence of non-NaCl compounds and crystals, in addition to the obvious decrease of NaCl purity and related loss of salty taste. Calcium sulphate is also a concern because of its insoluble nature and consequent precipitation in foods. For this reason, its presence should be diminished; actually, it is the main non-NaCl ionic compound normally associated with NaCl crystals (Heydarieh et al. 2020). Unrefined salt (diffused in developing Nations) is reported to contain 96% of NaCl, while remaining trace minerals are 4%. On the other side, refined salt is generally 99%-pure, but the presence of non-NaCl impurities is still reported (Heshmati et al. 2014; Hassan et al. 2017). From the viewpoint of food technologists and hygienists, the problem of microplastics (polyethylene, cellophane, polyethylene terephthalate, etc.) in salt from sea waters is a growing concern, as a result of environmental pollution. As a result, quality control and assurance issues are generally linked with salt, especially when speaking of commodities for the F&B industry. Contamination by foreign bodies (glass, wood, metals, etc.) in the food industry is a problem which has to be managed by means of the Hazard Analysis and Critical Control Points (HACCP) approach, posing serious challenges and the need of adequate inspections by means of metal and/or X-ray detectors, and proper HACCP training (Renzi et al. 2019; Yang et al. 2015). Anyway, each type of impurity causes some chemical and physical differences between different food-grade salts, with important commercial implications. As a simple example, a general colorimetric difference concerning Himalayan pink salt can be extremely important.

Consequently, edible salt types are generally correlated with easily detectable chemical and physical features. A possible classification taking into account notable differences can be shown as follows (Anjum et al. 2022; Goodall et al. 2000; O'Connor 2017; Powell and Miller 2018; Rodrigues et al. 2011; Saltwork Consultants 2022; Singh et al. 2023; Stergiou et al. 2016; Thomson and Wardle 1954):

(1) Table or common salts (97–99% of NaCl) contain sodium aluminosilicates or magnesium carbonates as anticaking agents (with a resulting flee-flowing effect). Iodine-fortified table salt is also available on the market. Depending on the original source (and also microbial populations in some area), these salts can be more or less coloured. Sea salts are generally more coloured than white table salts because of their content of non-sodium metals such as manganese, zinc, calcium, and potassium.

(2) Some of the most known salts obtained from sea waters are Fleur de sel, Grey salt, and Alaea salt (a sea salt mixed with red volcanic clay, produced in the Hawaii).

(3) Rock salts include the famous Himalayan pink salt (Khewra region, Pakistan) with different textures, and peculiar Persian blue salt (Semnan Province, Iran).

In the last situation, blue colour depends on the notable presence of sylvite and carnallite potassium minerals, causing a peculiar radiation damage.

(4) Dry, platy, or 'lamellose' crystals are generally named flaked salts (dendritic nature, with enhanced crunchiness and easy dissolution in water). Cyprian pyramid salt is a notable product with a pyramidal-crystal appearance. Kosher salt is a peculiar variation, being related to crystals large enough (this product is normally obtained from rock salt brines). River Murray salt flakes have a distinctive peach-like colour, obtained from continental surface brines (Australia).

(5) Smoked salts such as black *kala namak* (South East Asia), Hawaiian black salt, and *jugyeom* (South Korea) are normally wood-smoked or with activated charcoal, with specific tastes (activated charcoal gives a distinctive sulphur flavour).

As a result, it can be inferred that the main part of chemical and physical features with some impact on the F&B sector concern substantially:

(a) Water solubility.
(b) Salt crystal size and related texture and crunchiness.
(c) Colorimetric appearance, generally dependent on non-NaCl crystals (e.g. Persian blue salt), even if some processing method can alter it (smoked salt).
(d) The typical (mild or strong) taste, depending on non-sodium metals such as calcium and magnesium amounts in Australian peach-like salts.

The geographical localisation of certain salts is easily linked with sensorial features. With reference to certification systems, food-grade salts are generally offered on the market as 100%-natural salts, with the absence of genetically modified organisms, allergens, and gluten. Kosher certification is important, and the same thing is true when speaking of organic salt. Naturally, authenticity (food fraud issues) is required because of the strict link between geographical origin and processing methods on the one side, and sensorial/chemical and physical properties on the other side.

## 1.4 Milkfat in Food Products and Preparations. Basic Chemical Features

The category of 'milkfat' is variegated enough because of different reasons:

(a) The animal origin of milk (cow, buffalo, goat, camel, etc.).
(b) The geographical influence on the compositional features of obtained milkfat (also with reference to the use of different feeding materials).
(c) The used production processes for different milk-based products.

With reference to the basic aim of this book, the category of 'milkfat' comprehends only extracted lipids for subsequent food production. It has to be noted that several milk-based foods can be obtained from common milk(s) without the addition of milkfat, and/or from the mixture of different ingredients with the inclusion of milk-derived lipids. The discrimination of these products may be challenging enough from the commercial/economic viewpoint, and this fact has to be taken into account.

With concern to milkfats, a good starting point can be the Codex Standard for Milkfat Products. As mentioned in the Section No. 2 of this document, covered materials are essentially (Codex Alimentarius Commission 1973):

(1) Anhydrous milkfat.
(2) Milkfat.
(3) Anhydrous butteroil.
(4) Butteroil.
(5) Ghee.

Such a discrimination requires some explanations.

The first four sub-categories are mainly based on lipids of animal origin from milk(s) and/or milk products, and the extraction process has to be always intended as a removal of water and non-fat materials (Codex Alimentarius Commission 1973). Consequently, the original raw materials are milk(s) and/or milk products, and the production of milkfat matter is obtained by means of a separation process between fatty matters and non-fatty matters, including liquid water (which may generate water/oil emulsions). On the other side, 'ghee' is obtained from milk(s), milk cream, or milk butter (other milk products are excluded as raw materials), and the separation process allows for obtaining a specific texture and flavour for this fat material.

The chemical composition of these milkfat products should be investigated. For now, it should be noted that each final milkfat product may be produced with the use of lactic acid-microbial starters and with the addition of selected additives in accordance with the General Standard for Food Additives-CODEX STAN 192–1995 (Codex Alimentarius Commission 1995c), Food Category 02.1.1: ascorbyl esters and tocopherols (maximum allowed amount: 500 mg/kg), butylated hydroxyanisole (175 mg/kg), butylated hydroxytoluene (75 mg/kg), propyl gallate (100 mg/kg), citric acid, sodium dihydrogen citrate, and trisodium citrate (maximum allowed amount according to Good Manufacturing Practices). In addition, inert gases can be flushed before, during, and after filling operations into airtight containers, with the aim of avoiding rancidity because of oxidation reactions (Codex Alimentarius Commission 1973). For the remaining part, it should be taken into account that maximum water amount is allowed as 0.1% (weight/weight of total product) when speaking of anhydrous milkfat and butteroil (on the other side, fat matter should be $\geq$ 99.8%). Milkfat and butteroil without anhydrous claims should only have a minimum fat matter of 99.6% (weight/weight of total product), and the same thing is considered when speaking of ghee.

As a result, the name of the final product mixture has its own importance. In this ambit, the document: General Standard for the Use of Dairy Terms-CODEX STAN 206–1999 (Codex Alimentarius Commission 1999b) should be mandatorily considered before the discussion of basic chemical features concerning milkfat.

From a general viewpoint, 'milk' is the mammary secretion of milking animals (cow, buffalo, camel, etc.). The necessity of a notable amount or workable milk(s) implies that different milking animals of the same species are involved when speaking of milk collection, and the obtained milk(s) can be destined for consumption as liquid product (after adequate sanitisation treatments) or for subsequent food processing (Codex Alimentarius Commission 1999b).

On the other side, milk products are derived from milk processing treatments, including the addition of chemical compounds. It should be noted that a side category, composite milk products, are mainly obtained from milk, milk products, and/or separated milk constituents, on condition that non-milk-derived ingredients are not added with the specific aim of replacing totally or partially the role of milk constituents (in general terms: lactose and other carbohydrates; milkfat; mineral substances; milk casein and lactalbumins/lactoglobulins). As a result, such a composite milk product cannot be considered as raw material for milkfat production. Other dairy products (e.g. milk modified by means of addition or withdrawal of milk constituents) may be theoretically considered as basic raw materials for milkfat production (provided legal requirements are always met), but their use is not always rational. As an example, reconstituted milk products (obtained from water addition to dried or concentrated milk products) might be used as a raw material for milkfat separation; however, this operation is not clearly useful when speaking of milkfat production.

Consequently, the use of milk-related names is intended in a specific sense: the indication of animal origin of related liquid secretions. The use of milk-related names should not mislead food consumers, also with relation to obtained milkfat. As a simple example, 'butter' (maximum water amount: 16%; minimum fat amount: 80%) is related to milk origin only, while alternative names such as 'vegetable butter', 'plant butter', 'fruit butter', 'peanut butter', or 'butter blend' should be considered as misused options (Codex Alimentarius Commission 1999b). Butter, currently defined in the Codex Standard for Butter (CODEX STAN 279–1971), is among milkfat products: its importance on the economic ground is also linked to the fact that each similar product having > 95% can be named in this way, while selected additives (including sodium chloride) can be added (Codex Alimentarius Commission 1971).

After this long premise, some chemical and physical features concerning milkfat can be discussed. These facts are presented with reference to already separated milkfat, while milkfat in its native form is not discussed here.

Basically, and with specific reference to cow milk only (the most used raw material in this ambit), 3.25–3.7% of liquid milk corresponds to the fat portion (while water is 87% and total solids are approximately 13%). This organic phase is able to carry other important molecules, such as vitamins A, D, E, and K (these compounds are all fat-soluble molecules). The importance of fat matter in milk is notable: in addition, the design of many food products relies on milkfat presence. For example, and with concern to milk-derived products only, the following foods have to be considered in this ambit, with the obvious exclusion of original milk (IDFA 2024; Jensen et al. 1991):

(1)  Cultured milk.
(2)  Milk & cream mixture (approximately 1:1), while milkfat should be comprised between 10.5 and 18%.
(3)  Light cream (milkfat is comprised between 18 and 30%).
(4)  Light whipping cream (milkfat is comprised between 30 and 36%).
(5)  Heavy whipping cream (milkfat is > 36%).
(6)  Sour cream (it is a pasteurised cream with lactic acid bacteria culture, and milkfat is ≥ 18%).
(7)  Yogurt (milkfat has to be ≥ 3.25%).
(8)  Ice cream (in the United States of America, milkfat has to be ≥ 10%).
(9)  Different cheeses, where fat content can be extremely variable.

When speaking of the fat phase only, approximately 93–98% of the material is composed of triacylglycerol, while phospholipids (membrane stabilisers in the original milk) and sterols (mainly cholesterol) do not normally exceed 1.0 and 0.5%, respectively (Deeth 1997; Huppertz et al. 2020; Lopez 2005; Mohan et al. 2021). Triglycerides (a general structure for a saturated triacylglycerol is shown in Fig. 1.7, taking into account normal molecules are unsaturated fat triglycerides) are based on a $C_{4:0}$ to $C_{10:0}$ structure per each aliphatic chain with important $C_{18:2}$ and *trans*-$C_{18:1}$ portions. After separation (by centrifugation, as a single example), the lipid phase loses its globular form. In addition, and this fact has to be taken seriously into account, lipids are not protected from oxidative reactions with consequent lipolysis and rancid odours. In this situation, the composition is not modified when speaking of fatty acids, but the quality and resistance of fat matter are seriously compromised. As a single example, butyric flavour is always associated with rancidity (Jensen et al. 1991).

With reference to industrial uses in the food industry, one of the most interesting—and challenging—properties of milkfat is attitude to crystallisation. In other terms,

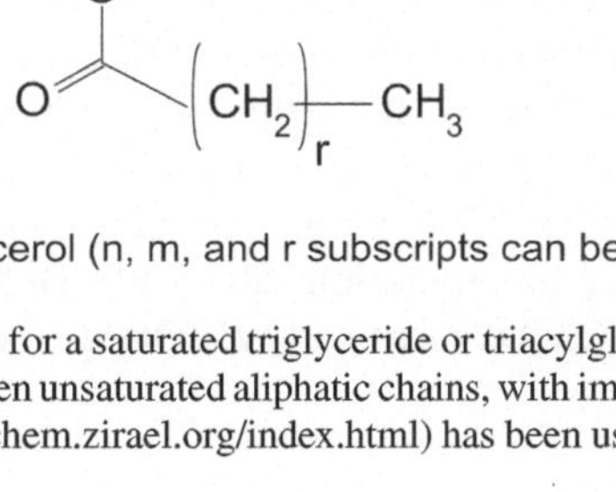

A saturated triacylglycerol (n, m, and r subscripts can be different numbers)

**Fig. 1.7**  A simplified structure for a saturated triglyceride or triacylglycerol. It has to be noted that natural triglycerides contain often unsaturated aliphatic chains, with important differences. BKchem version 0.13.0, 2009 (http://bkchem.zirael.org/index.html) has been used for drawing this structure

the application of milkfat as an additive has to be considered with care because of the tendency to crystallisation with consequent hardness and cloudiness. The need to improve spreadability of food preparations, where possible, has been always a challenging problem. Monoacylglycerols and other emulsifying agents have been used, such as other polar lipids, with the aim of stabilising preferred β′ crystal polymorphs instead of the β form which is thermodynamically more stable and undesirable at the same time (Mohan et al. 2021; Viriato et al. 2018; Waldron et al. 2020).

This behaviour of milkfat products depends also on the variability of its lipid composition with consequent opportunities when speaking of milkfat fractionation and cholesterol removal (Hogan and O'Callaghan 2020). Being butter and other milkfat products plastic emulsions with solidification behaviour at room temperature (water/oil emulsions), the main efforts of food technologists have been to homogenise globule sizes with consequent dispersion of water droplets in a continuous fat network made of little β′ crystal polymorphs (Keogh 2006). At the same time, the relatively notable amount of saturated fatty acids and cholesterol should be advice to attempt fractionation with the aim of obtaining nutritionally favourable and healthy recommendable animal fat matters (Chen et al. 1992).

Consequently, the efforts of the food industry are mainly linked to the below-mentioned topics:

(a) Production of highly homogenisable milkfat (with ameliorated rheological properties).
(b) Production of fat agglomerates in peculiar applications (butter, whipping, and ice creams) with the aim of promoting coalescence (clustering and clumping of fat globules) in a reversible or irreversible way, depending on applications.
(c) Production of nutritionally recommendable milkfat fractions (limited cholesterol amount, improved separation of saturated fatty acids, inhibition of butyric rancidity, etc.).
(d) Production of milkfat where long-chain fatty acids are prevailing (with associated excellent flavour). The following factors are highly desirable: useful effects on protein profiles in cheese (with reference to good separation between nitrogen-containing chains into the cheese network); interesting colours (fat matter is also a carrier for natural carotenoids); improved rheology (butter has a 16–24 °C thermal interval for optimal spreadability).

Finally, it should be remembered that fat matter has a distinctive role in cheeses when speaking of the limiting effect of water absorption by casein in cheese as a function of fat matter on dry content. In other terms, the higher the fat matter on dry content (often increased by incorporation of milkfat), the lower the water absorbtion by casein in cheese (Parisi 2012). This result has a certain impact on cheesemaking yields which can be calculated by means of an indirect system for the determination of protein amount in cheeses. This system has been also inserted into two patented software products: 'Water Incorporation, Shelf–life Deduction and Orientation in the Modeling of' (WISDOM) Cheese and WISDOM Cheese II (Parisi et al. 2019b, 2020).

## 1.5   Sugar, Salt, and Milkfat in Food Products and Preparations. A Joint Chemical Perspective in Terms of Flavour and Satiety Ingredients

The above-mentioned chemical and physical features for salt, sugars, and milkfat have important implications when speaking of edible foods and beverages. This section discusses in brief each of these categories with reference to specific characteristics linked to palatability and satiety performances.

### 1.5.1   Sugars in Food Preparations. Functional Features, Taste, and Satiety

Taking into account chemical and physical features discussed in Sect. 1.2, sugars have many functions (if compared with NaCl) when speaking of food preparations (EUFIC 2024; Koivistoinen and Hyvönen 1985; IFST 2022; IFT 2019):

(a)  Sweetening power, with possible synergic effects in some products such as chocolates (bitterness perception may be positively influenced; in addition, too acidic tastes may be modified positively, such as in tomato sauces and other fluid foods).

(b)  Colorimetric enhancement/browning as the result of Maillard reactions. This result can be extremely interesting in certain products such as baked foods, while the same products cannot optically be accepted in other products such as dairy foods (Martins et al. 2000; Parisi and Luo 2018; Parisi et al. 2019a; Singla et al. 2018; Vaclavik and Christian 2014a,b).

(c)  Flavour improvement by means of synergic effects.

(d)  Textural enhancement with associated volumetric increase and/or lightness modifications where possible (interested foods can be baked foods, jams, frozen desserts, etc.). The improvement of crystallisation and reduction of freezing point in ice creams (with associated milkfat), or increase of boiling point in the production of candies have to be well considered. The association of sugars with gelling agents (pectins) is extremely important in jams.

(e)  Positive influence on microbial ecology. Sugars can be microbially fermented (leavened breads, etc.); however, related addition can be also strategically used to inhibit microbial spreading because of water-binding power (similarly to NaCl), with associated humectant effect. Osmotic pressure is notably increased. As a collateral result, shelf life can be extended, even it degradation is always inevitable, according to the First Parisi's Law of Food Degradation (Anonymous 2021; Parisi 2002).

With concern to the increasing social problem of obesity, it should be noted that sugars are reported to stimulate satiety (intended as a temporal period which means clearly eating cessation, while a similar concept—satiation—is for 'intra-meal satiety'). As a result, short-term food intake is generally reduced (Anderson and Woodend 2003). This fact has been also reported in comparison with artificial sweeteners: in detail, these artificial compounds have minimal effects on appetite, while sugars are able to stimulate satiety in terms of suppression of ghrelin hormone in the gastrointestinal tract (Steinert et al. 2011).

## 1.5.2  Salt in Food Preparations. Functional and Taste Features

Basically, and from the historical viewpoint, salt has always had three main functions when speaking of food preparations (Man 2007):

(a) Salty tastes, with possible enhancement of other flavour components in a deter-mined food. Saltiness perceptions depend on different factors, including salt and aroma release, and secondary perception mechanisms (gustation and olfac-tion) in the human oro-nasal cavity. As a result, the human brain integrated different stimuli with a multisensorial resulting perception, also indicating synergic action from non-salt substances. This reflection can be useful when speaking of salt reduction, as recently demonstrated with reference to umami-related substances (sodium glutamate and disodium inosinate) and their ability to enhance perceived saltiness. Some peptides and other molecules can be studied in this ambit, taking into account that generally salt can enhance the perceived perception of acids and vice versa, provided that concentrations are not high (otherwise, suppression of perceived tastes should be expected). In addition, salt can possibly suppress bitter taste while bitter compounds have no evident (suppressive) effects on saltiness perception (Breslin 1996; Li et al. 2024; Sahni et al. 2023; Thomas-Danguin et al. 2019).

(b) Secondly, food processing and manipulation can be influenced depending on salt content. One of the most recent sanitising systems in food industries is the use of electrolysed oxidising water because of (i) the needed use of water and NaCl only, and (ii) related sanitising advantages (powerful disinfection results; friendly-user performance, including no corrosion dangers for human operators and stainless surfaces). In addition, such a system would be 'green' enough to justify related environmental claims, while the most important disadvantage is the continuous electrolysis (otherwise, antimicrobial power would rapidly decrease) (Huang et al. 2008; Rebezov et al. 2022). Salt is generally used in many cheese preparations, including processed cheeses. In addition to preservative reasons, NaCl is also able to bind a certain water amount, and this fact should be taken into account when speaking of concentrated or dehydrated foods (Henning et al. 2006). On the other side, excessive salt amounts could give excessive water losses if undesired with consequent handling and storage risks such as remarkable hygroscopicity.

(c)  Preservation properties have a key and historical importance when speaking of salt in foods (antimicrobial features because of low water activity values with consequent osmotic effects and reduction of general water content in certain foods). Main results can be explained in this way. On the other side, it has to be noted that preservative processing cannot only rely on salt: other different systems are required with a synergic action. The well-known water-binding action depends also on protein-related features (capillary sizes in certain preparations, the possibility of hydrophobic interactions and disulphide bonds in certain protein agglomerations, presence of different ionic atoms, preferential bonding between chloride ion and protein, pH below isoelectric point, etc.). As a result, dehydration occurs generally if salt addition is remarkable because NaCl and other solutes compete with protein for the chemically bonding-available water (Albarracín et al. 2011). For these reasons, salt is a really useful preservation system but it should be always used in a broader ambit (more than one single preservation or dehydration strategy would be recommended, in general).

### 1.5.3  Milkfat in Food Preparations. Functional and Taste Features

As above mentioned, milkfat products have the following main functions when speaking of food preparations:

(a)  Improvement of rheology in certain products.
(b)  Promotion of coalescence in a reversible or irreversible way, depending on final desired products.
(c)  Improved resistance to lipid oxidation with associated increase of durability.
(d)  Enhancement of flavour.
(e)  Amelioration of cheese structure (also by means of the limitation of water absorption by protein molecules), colour, spreadability, etc.

Some final note can be added with reference to satiety. In particular (Abou Samra 2010; Erlanson-Albertsson 2010; Pangborn et al. 1985), satiety is positively influenced if the amount of long-chain triacylglycerols is lower that short-chain molecules. In detail, gut hormone secretion appears to be dependent on chain length. Consequently, should long-chain molecules be abundant enough, energy intake would be easily suppressed. On the other side, medium-chain triacylglycerols can cause nausea, gastrointestinal disorders, etc.; as a consequence, the intake of large amounts of medium-chain fat molecules is generally inhibited. High unsaturation and esterification degrees have been also linked with enhanced satiety, but more research is needed at present. It has been reported that melting point of fat materials has to be some effect on unsaturation degree, with consequent effects on satiety. In addition, fatty acid oxidation in the liver should have some importance when speaking of energy intake suppression. Surely, enhanced satiety can be induced when fibre is associated with fat consumption.

# References

Abou Samra R (2010) Fats and satiety. In: Montmayeur JP, le Coutre J (eds) Fat detection: taste, texture, and post ingestive effects. CRC Press/Taylor & Francis, Boca Raton, FL

Ackroff K, Sclafani A (1999) Palatability and foraging cost interact to control caloric intake. J Exp Psycho Anim Behav Proc 25(1):28–36. https://doi.org/10.1037/0097-7403.25.1.28

Albarracín W, Sánchez IC, Grau R, Barat JM (2011) Salt in food processing; usage and reduction: a review. Int J Food Sci Technol 46(7):1329–1336. https://doi.org/10.1111/j.1365-2621.2010.02492.x

Allstun T (2013) A sweet (or not so sweet) surprise: unpacking the FDA ruling against corn sugar as an alternative name for high-fructose corn syrup. Drake J Agric Law 18(2):349–374

Anderson GH, Woodend D (2003) Consumption of sugars and the regulation of short-term satiety and food intake. Am J Clin Nutr 78(4):843S–849S. https://doi.org/10.1093/ajcn/78.4.843S

Anjum MI, ur Rehman S, Kakakhel MB, Siddique MT, Mahmood MM, Hayat S, Ahmad K (2022) Thermoluminescence study of Pink Himalayan salt from Khewra mines, Pakistan. J Lumin 252:119329. https://doi.org/10.1016/j.jlumin.2022.119329

Anonymous (2021) Parisi's first law of food degradation valuable to establish adequate protocols concerning food durability. Inside Lab Manag 25(1):17. AOAC International, Rockville, MD

Baglio E (2014) Chemistry and technology of yoghurt fermentation. Springer International Publishing, Cham

Baglio E (2017) Chemistry and technology of honey production. Springer International Publishing, Cham

Barbieri G, Barone C, Bhagat A, Caruso G, Conley ZR, Parisi S (2014) The influence of chemistry on new foods and traditional products. Springer International Publishing, Cham

Barone C, Barbera M, Barone M, Parisi S, Zaccheo A (2017) Chemical evolution of nitrogen-based compounds in mozzarella cheeses. Springer International Publishing, Cham

Belc N, Smeu I, Macri A, Vallauri D, Flynn K (2019) Reformulating foods to meet current scientific knowledge about salt, sugar and fats. Trends Food Sci Technol 84:25–28. https://doi.org/10.1016/j.tifs.2018.11.002

Bozkaya O, Aluç Y (2024) Physico-chemical characterization of food grade natural spring salt from the Central Anatolia region of Turkey and investigation of its microplastic content. J Food Sci Technol 2024. https://doi.org/10.1007/s13197-024-05942-0

Breslin PA (1996) Interactions among salty, sour and bitter compounds. Trends Food Sci Technol 7(12):390–399. https://doi.org/10.1016/S0924-2244(96)10039-X

Chen H, Schwartz SJ, Spanos GA (1992) Fractionation of butter oil by supercritical carbon dioxide. J Dairy Sci 75(10):2659–2669. https://doi.org/10.3168/jds.S0022-0302(92)78027-8

Clemens RA, Jones JM, Kern M, Lee SY, Mayhew EJ, Slavin JL, Zivanovic S (2016) Functionality of sugars in foods and health. Compr Rev Food Sci Food Saf 15(3):433–470. https://doi.org/10.1111/1541-4337.12194

Codex Alimentarius Commission (1971) Codex standard for butter (CODEX STAN 279–1971). The food and agriculture organization of the United Nations (FAO), Rome, and the World Health Organization (WHO), Geneva

Codex Alimentarius Commission (1973) Codex standard for milkfat products (CODEX STAN 280–1973). The food and agriculture organization of the United Nations (FAO), Rome, and the World Health Organization (WHO), Geneva

Codex Alimentarius Commission (1995a) Codex standard for food grade salt (CODEX STAN 150–1995). The food and agriculture organization of the United Nations (FAO), Rome, and the World Health Organization (WHO), Geneva

Codex Alimentarius Commission (1995b) Codex general standard for contaminants and toxins in foods and feeds (CODEX STAN 193–1995). The food and agriculture organization of the United Nations (FAO), Rome, and the World Health Organization (WHO), Geneva

Codex Alimentarius Commission (1995c) General standard for food additives (CODEX STAN 192–1995). The food and agriculture organization of the United Nations (FAO), Rome, and the World Health Organization (WHO), Geneva

Codex Alimentarius Commission (1999a) Standard for sugars (CXS 212–1999). The food and agriculture organization of the United Nations (FAO), Rome, and the World Health Organization (WHO), Geneva

Codex Alimentarius Commission (1999b) General standard for the use of dairy terms (CODEX STAN 206–1999). The food and agriculture organization of the United Nations (FAO), Rome, and the World Health Organization (WHO), Geneva

Coultate TP (2009) Food: the chemistry of its components, 5th edn. The Royal Society of Chemistry, Cambridge

Davis EA (1995) Functionality of sugars: physicochemical interactions in foods. Am J Clin Nutr 62(1):170S-177S. https://doi.org/10.1093/ajcn/62.1.170S

Deeth HC (1997) The role of phospholipids in the stability of milk fat goblules. Aust J Dairy Technol 52:44–46

Delahunty CM, Havermans RC (2010) The sensory systems and food palatability. In: Roche HM, Macdonald IA, Schols AMWJ, Lamham-New SA (eds) Nutrition and metabolism, 3rd edn. Wiley, Hoboken and Chichester

Drewnowski A, Monsivais P (2020) Taste, cost, convenience, and food choices. In: Marriott BP, Birt DF, Stallings VA, Yates AA (eds) Present knowledge in nutrition, volume 2: clinical and applied topics in nutrition, 11th edn. Academic Press, London, San Diego, Cambridge, and Oxford, pp 185–200. https://doi.org/10.1016/B978-0-12-818460-8.00010-1

Erlanson-Albertsson C (2010) Fat-rich food palatability and appetite regulation. In: Montmayeur JP, le Coutre J (eds) Fat detection: taste, texture, and post ingestive effects. CRC Press/Taylor & Francis, Boca Raton, FL

EUFIC (2024) What is the role of sugar in the food industry? European food information council (EUFIC), Brussels, https://www.eufic.org/en/. Available https://www.eufic.org/en/whats-in-food/article/sugars-from-a-food-technology-perspective. Accessed 28th May 2024

Feofilova EP, Usov AI, Mysyakina IS, Kochkina GA (2014) Trehalose: chemical structure, biological functions, and practical application. Microbiol 83:184–194. https://doi.org/10.1134/S0026261714020064

Fulton A, Charles D (2012) FDA rules corn syrup can't change its name to corn sugar. National public radio (NPR), Washington, D.C., www.npr.org. Available https://www.npr.org/sections/thesalt/2012/05/30/154009682/fda-rules-corn-syrup-cant-change-its-name-to-corn-sugar. Accessed 28 May 2024

Gates P, Copeland J, Stevenson RJ, Dillon P (2007) The influence of product packaging on young people's palatability rating for RTDs and other alcoholic beverages. Alcohol Alcohol 42(2):138–142. https://doi.org/10.1093/alcalc/agl113

Gibson S, Ashwell M (2011) Dietary patterns among British adults: compatibility with dietary guidelines for salt/sodium, fat, saturated fat and sugars. Pub Health Nutr 14(8):1323–1336. https://doi.org/10.1017/S1368980011000875

Goodall TM, North CP, Glennie KW (2000) Surface and subsurface sedimentary structures produced by salt crusts. Sedimentol 47(1):99–118. https://doi.org/10.1046/j.1365-3091.2000.00279.x

Greenwald RP, Childs L, Pachón H, Timmer A, Houston R, Codling K (2022) Comparison of salt iodization requirements in national standards with global guidelines. Curr Develop Nutr 6(8):116. https://doi.org/10.1093/cdn/nzac116

Ha OR, Killian H, Bruce JM, Lim SL, Bruce AS (2018) Food advertising literacy training reduces the importance of taste in children's food decision-making: a pilot study. Front Psychol 9:374895. https://doi.org/10.3389/fpsyg.2018.01293

Haley S, Suarez N (2004) Sugar and sweetener situation and outlook yearbook, No. SSS-2004, US Department of Agriculture, Economic Research Service (ERS-NASS), Washington, D.C., p 81. https://doi.org/10.5555/20043188692

Henning DR, Baer RJ, Hassan AN, Dave R (2006) Major advances in concentrated and dry milk products, cheese, and milk fat-based spreads. J Dairy Sci 89(4):1179–1188. https://doi.org/10.3168/jds.S0022-0302(06)72187-7

Heshmati A, Vahidinia A, Salehi I (2014) Evaluation of heavy metals contamination of unrefined and refined table salt. Int J Res Stud Biosci 2(2):21–24

Heydarieh A, Arabameri M, Ebrahimi A, As'habi A, Marvdashti LM, Yanche-shmeh BS, Abdol-shahi A (2020) Determination of magnesium, calcium and sulphate Ion impurities in commercial edible salt. J Chem Health Risks 10(2):93–102. https://doi.org/10.22034/jchr.2020.1883343.1067

Hoffmann CM (2010) Root quality of sugarbeet. Sugar Tech 12(3):276–287. https://doi.org/10.1007/s12355-010-0040-6

Hogan SA, O'Callaghan TF (2020) Milk fat: chemical and physical modification. In: McSweeney PLH, Fox PF, O'Mahony JA (eds) Advanced dairy chemistry volume 2 lipids, pp 197–217. https://doi.org/10.1007/978-3-030-48686-0_7

Hsu T (2012) FDA rejects bid to rename high-fructose corn syrup corn sugar. Los Angeles Times, Los Angeles, CA. Available https://www.latimes.com/business/la-fi-mo-fda-corn-sugar-20120531-story.html. Accessed 28 May 2024

Huang YR, Hung YC, Hsu SY, Huang YW, Hwang DF (2008) Application of electrolyzed water in the food industry. Food Control 19(4):329–345. https://doi.org/10.1016/j.foodcont.2007.08.012

Huppertz T, Uniacke-Lowe T, Kelly AL (2020) Physical chemistry of milk fat globules. In: McSweeney PLH, Fox PF, O'Mahony JA (eds) Advanced dairy chemistry volume 2 lipids, pp 133–167. https://doi.org/10.1007/978-3-030-48686-0_5

IDFA (2024) Definitions. International dairy foods association (IDFA). Washington, D.C., www.idfa.org. Available https://www.idfa.org/definition. Accessed 28 May 2024

IFST (2022) Sugars. Institute of food science & technology (IFST). London. Available https://www.ifst.org/resources/information-statements/sugars. Accessed 28 May 2024

IFT (2019) Sugars: a scientific overview. Institute of food technologists (IFT). Chicago. Available https://www.ift.org/career-development/learn-about-food-science/food-facts/food-facts-food-ingredients-and-additives/sugars-a-scientific-overview. Accessed 28 May 2024

Jensen RG, Ferris AM, Lammi-Keefe CJ (1991) The composition of milk fat. J Dairy Sci 74(9):3228–3243. https://doi.org/10.3168/jds.S0022-0302(91)78509-3

Just DR, Sığırcı Ö, Wansink B (2014) Lower buffet prices lead to less taste satisfaction. J Sens Stud 29(5):362–370. https://doi.org/10.1111/joss.12117

Keogh MK (2006) Chemistry and technology of butter and milk fat spreads. In: Fox PF, McSweeney PLH (eds) Advanced dairy chemistry volume 2 lipids, pp 333–363. https://doi.org/10.1007/0-387-28813-9_9

Khandpur N, Rimm EB, Moran AJ (2020) The influence of the new US nutrition facts label on consumer perceptions and understanding of added sugars: a randomized controlled experiment. J Acad Nutr Diet 120(2):197–209. https://doi.org/10.1016/j.jand.2019.10.008

Kim JY, Prescott J, Kim KO (2014) Patterns of sweet liking in sucrose solutions and beverages. Food Qual Pref 36:96–103. https://doi.org/10.1016/j.foodqual.2014.03.009

Kim EJ, Ellison B, McFadden B, Prescott MP (2021) Consumers decisions to access or avoid added sugars information on the updated nutrition facts label. PLoS ONE 16(3):e0249355. https://doi.org/10.1371/journal.pone.0249355

Klosse PR, Riga J, Cramwinckel AB, Saris WH (2004) The formulation and evaluation of culinary success factors (CSFs) that determine the palatability of food. Food Serv Technol 4(3):107–115. https://doi.org/10.1111/j.1471-5740.2004.00097.x

Koivistoinen P, Hyvönen L (1985) The use of sugar in foods. Int Dent J 35(3):175–179

Kyle TK, Thomas DM (2014) Consumers believe nutrition facts labeling for added sugar will be more helpful than confusing. Obes 22(12):2481–2484. https://doi.org/10.1002/oby.20887

Lambert EG, O'Keeffe CJ, Ward AO, Anderson TA, Yip Q, Newman PL (2024) Enhancing the palatability of cultivated meat. Trends Biotechnol 24:2024. https://doi.org/10.1016/j.tibtech.2024.02.014

Li J, Zhong F, Spence C, Xia Y (2024) Synergistic effect of combining umami substances enhances perceived saltiness. Food Res Int 189:114516. https://doi.org/10.1016/j.foodres.2024.114516

Lillford PJ (2016) The impact of food structure on taste and digestibility. Food Funct 7(10):4131–4136. https://doi.org/10.1039/C5FO01375E

Lopez C (2005) Focus on the supramolecular structure of milk fat in dairy products. Reprod Nutr Develop 45(4):497–511. https://doi.org/10.1051/rnd:2005034

Low YQ, Lacy K, Keast R (2014) The role of sweet taste in satiation and satiety. Nutr 6(9):3431–3450. https://doi.org/10.3390/nu6093431

Mahato DK, Keast R, Liem DG, Russell CG, Cicerale S, Gamlath S (2020) Sugar reduction in dairy food: an overview with flavoured milk as an example. Foods 9(10):1400. https://doi.org/10.3390/foods9101400

Man CMD (2007) Technological functions of salt in food products. In: Kilcast D, Angus F (Eds) (2007) Reducing salt in foods: practical strategies. Woodhead Publishing Ltd. Cambridge, pp 157–173. https://doi.org/10.1533/9781845693046.2.157

Martins SIFS, Jongen WMF, Van Boekel MAJS (2000) A review of maillard reaction in food and implications to kinetic modelling. Trends Food Sci Technol 11(9–10):364–373. https://doi.org/10.1016/S0924-2244(01)00022-X

Mohan MS, O'Callaghan TF, Kelly P, Hogan SA (2021) Milk fat: opportunities, challenges and innovation. Crit Rev Food Sci Nutr 61(14):2411–2443. https://doi.org/10.1080/10408398.2020.1778631

Mukhopadhyay S, Goswami S, Mondal SA, Dutta D (2020) Dietary fat, salt, and sugar: a clinical perspective of the social catastrophe. In: Preuss HG, Bagchi D (eds) Dietary sugar, salt and fat in human health. Academic Press, London, San Diego, Cambridge, and Oxford, pp 67–91. https://doi.org/10.1016/B978-0-12-816918-6.00003-2

O'Connor K (2017) Seaweed: a global history. Reaktion Books, London

Pangborn RM, Bos KE, Stern JS (1985) Dietary fat intake and taste responses to fat in milk by under-, normal, and overweight women. Appet 6(1):25–40. https://doi.org/10.1016/S0195-6663(85)80048-9

Parisi S (2002) Profili evolutivi dei contenuti batterici e chimico-fisici in prodotti lattiero-caseari. Ind Aliment 412(41):295–306

Parisi S (2012) La mutua ripartizione tra lipidi e caseine nei formaggi Un Approccio Simulato. Ind Aliment 523(51):7–15

Parisi S, Ameen SM, Montalto S, Santangelo A (2019a) Maillard reaction in foods. Springer International Publishing, Cham

Parisi S, Barone C, Laganà P (2019b) WISDOM (Water incorporation, shelf–life deduction and orientation in the modeling of) cheese. Italian Patent (demand No. 102017000041086), date: 27 Sept 2019

Parisi S, Barone C, Laganà P (2020) WISDOM (Water incorporation, shelf–life deduction and orientation in the modeling of) Cheese II. Italian patent (demand No. 102018000004178), date: 08 April 2020

Parisi S, Luo W (2018) The importance of maillard reaction in processed foods. Springer International Publishing, Cham

Patterson NJ, Sadler MJ, Cooper JM (2012) Consumer understanding of sugars claims on food and drink products. Nutr Bull 37(2):121–130. https://doi.org/10.1111/j.1467-3010.2012.01958.x

Powell WK, Miller PL (2018) Hawaiian shamanistic healing: medicine ways to cultivate the aloha spirit. Llewellyn Worldwide, Woodbury, MN

Rebezov M, Saeed K, Khaliq A, Rahman SJU, Sameed N, Semenova A, Khayrullin M, Dydykin A, Abramov Y, Thiruvengadam M, Shariati MA, Bangar SP, Lorenzo JM (2022) Application of electrolyzed water in the food industry: a review. Appl Sci 12(13):6639. https://doi.org/10.3390/app12136639

Renzi M, Grazioli E, Bertacchini E, Blašković A (2019) Microparticles in table salt: levels and chemical composition of the smallest dimensional fraction. J Mar Sci Eng 7(9):310. https://doi.org/10.3390/jmse7090310

Rodrigues CM, Bio A, Amat F, Vieira N (2011) Artisanal salt production in Aveiro/Portugal-an ecofriendly process. Saline Syst 7(3):14. https://doi.org/10.1186/1746-1448-7-3

Sahni O, Didzbalis J, Munafo JP Jr (2023) Saltiness enhancement through the synergism of pyroglutamyl peptides and organic acids. J Agric Food Chem 72(1):625–633. https://doi.org/10.1021/acs.jafc.3c05911

Saltwork Consultants (2022) Artisanal salt and culinary expectations. Saltwork Consultants Pty Ltd., Kingston Park. Available https://www.saltworkconsultants.com/culinary-salt/. Accessed 28 May 2024

Scapin T, Fernandes AC, Proença RPDC (2017) Added sugars: definitions, classifications, metabolism and health implications. Rev Nutr 30:663–677. https://doi.org/10.1590/1678-986 52017000500011

Schorin MD, Sollid K, Edge MS, Bouchoux A (2012) The science of sugars, Part I: a closer look at sugars. Nutr Today 47(3):96–101. https://doi.org/10.1097/NT.0b013e3182435de8

Shakoor A, Zhang C, Xie J, Yang X (2022) Maillard reaction chemistry in formation of critical intermediates and flavour compounds and their antioxidant properties. Food Chem 393:133416. https://doi.org/10.1016/j.foodchem.2022.133416

Sigman-Grant M, Morita J (2003) Defining and interpreting intakes of sugars. Am J Clin Nutr 78(4):815S–826S. https://doi.org/10.1093/ajcn/78.4.815S

Singh K, Tyagi SK, Chaudhary M, Baliyan D (2023) Traditional and rural manufacturing process of black salt. Pharm Innov J 12(6):1977–1980

Singla RK, Dubey AK, Ameen SM, Montalto S, Parisi S (2018) Analytical methods for the assessment of maillard reactions in foods. Springer International Publishing, Cham

Starkey DE, Wang Z, Brunt K, Dreyfuss L, Haselberger PA, Holroyd SE, Janakiraman K, Kasturi P, Konings EJM, Labbe D, Latulippe ME, Lavigne X, McCleary BV, Parisi S, Shao T, Sullivan D, Torres M, Yadlapalli S, Vrasidas I (2022) The challenge of measuring sweet taste in food ingredients and products for regulatory compliance: a scientific opinion. J AOAC Int 105(2):333–345. https://doi.org/10.1093/jaoacint/qsac005

Steinert RE, Frey F, Töpfer A, Drewe J, Beglinger C (2011) Effects of carbohydrate sugars and artificial sweeteners on appetite and the secretion of gastrointestinal satiety peptides. Brit J Nutr 105(9):1320–1328. https://doi.org/10.1017/S000711451000512X

Stergiou C, Karageorgiou S, Theodoridou S, Giouri K, Papadopoulou L, Melfos V (2016) Compositional and morphological evaluation of edible salts: preliminary results. Bull Geol Soc Greece 50(4):2018–2024. https://doi.org/10.12681/bgsg.11948

Stylianopoulos C (2013) Carbohydrates: chemistry and classification. In: Caballero B (ed) Encyclopedia of human nutrition, 3rd edn. Academic Press, London, San Diego, Cambridge, and Oxford. https://doi.org/10.1016/B978-0-12-375083-9.00041-6

Su J (2000) Sucrose, the most popular sweet food and food sweetener. Food Sci Agric Chem 2(1):1–6. https://doi.org/10.5555/20000307088

Tappy L, Lê KA (2010) Metabolic effects of fructose and the worldwide increase in obesity. Physiol Rev 90(1):23–46. https://doi.org/10.1152/physrev.00019.2009

Thomas-Danguin T, Guichard E, Salles C (2019) Cross-modal interactions as a strategy to enhance salty taste and to maintain liking of low-salt food: a review. Food Funct 10(9):5269–5281. https://doi.org/10.1039/C8FO02006J

Thomson SJ, Wardle G (1954) Coloured natural rocksalts: a study of their helium contents, colours and impurities. Geochim Cosmochim Acta 5(4):169–184. https://doi.org/10.1016/0016-703 7(54)90031-9

Twarda-Clapa A, Olczak A, Białkowska AM, Koziołkiewicz M (2022) Advanced glycation endproducts (AGEs): formation, chemistry, classification, receptors, and diseases related to AGEs. Cells 11(8):1312. https://doi.org/10.3390/cells11081312

ul Hassan A, Din AMU, Ali S (2017) Chemical characterisation of Himalayan rock salt. Pak J Sci Ind Res Ser A: Phys Sci 60(2):67–71

Vaclavik V, Christian E (2014a) Carbohydrates in food. In: Heldman D (ed) Essentials of food science, 4th edn. Springer International Publishing, New York, NY, pp 27–37

Vaclavik V, Christian E (2014b) Sugars, sweeteners. In: Heldman D (ed) Essentials of food science, 4th edn. Springer International Publishing, New York, NY, pp 279–295

Viriato RLS, de Souza QM, da Gama MAS, Ribeiro APB, Gigante ML (2018) Milk fat as a structuring agent of plastic lipid bases. Food Res Int 111:120–129. https://doi.org/10.1016/j.foo dres.2018.05.015

Waldron DS, Hoffmann W, Buchheim W, McMahon DJ, Goff HD, Crowley SV, Moloney C, O'Regan J, Giuffrida F, Celigueta Torres I, Siong P (2020) Role of milk fat in dairy products. In: McSweeney PLH, Fox PF, O'Mahony JA (eds) Advanced dairy chemistry volume 2 lipids, pp 245–305. https://doi.org/10.1007/978-3-030-48686-0_9

Wrolstad RE (2012) Food carbohydrate chemistry, vol 48. Wiley-Blackwell, Chichester, and Institute of Food Technologists Press, Chicago

Yang D, Shi H, Li L, Li J, Jabeen K, Kolandhasamy P (2015) Microplastic pollution in table salts from China. Environ Sci Technol 49(22):13622–13627. https://doi.org/10.1021/acs.est.5b03163

Yeomans MR (1998) Taste, palatability and the control of appetite. Proc Nutr Soc 57:609–615. https://doi.org/10.1079/PNS19980089

Young JS (2021) Measuring palatability as a linear combination of nutrient levels in food items. Food Policy 104:102146. https://doi.org/10.1016/j.foodpol.2021.102146

# Chapter 2
# Advertising of Unhealthy Foods from a Scientific Perspective

**Abstract** With reference to the sector of food and beverage preparations, three nutritional categories and ingredients—sugars, salt, and milkfat—have been and still are used for enhancing taste, palatability, and satiety, with a few exceptions only when speaking of final products. However, current marketing and advertising strategies are oriented towards the representation of foods containing low-calorie sweeteners, low-sodium salts, and unrefined vegetable oils. The main reason is that these typologies are represented as unhealthy food ingredients with a strong correlation between excessive intake and different diseases. The aim of this chapter is to discuss advertising-related aspects for each of the three above-mentioned food categories: sugars, salt, and milkfat. In general, the increase of sold units with consequent decrease of production costs is directly linked to technical features and commercial factors at the same time. Secondly, the conversion of technical and non-technical information on food labels in a practical and comprehensible marketing/advertising way for food and beverage consumers is not a simple work. Certain problems should be avoided with a careful design (promotion of useful factors, elimination of difficult concepts). Moreover, the continuous improvement of food and beverage products and services from time to time depends on the success on the market and also on the attention of the consumer. Nutrition and health claims for sugars, salt, and milkfat are studied in detail with the aim of clarifying the current situation for 'unhealthy foods'.

**Keywords** Health claim · Non-communicable disease · Nutrition claim · Salt · Saturated fat · Sugar · Unhealthy food

## Abbreviations

| | |
|---|---|
| EU | European Union |
| F&B | Food and beverage |
| FBO | Food business operator |
| GFSI | Global Food Safety Initiative |
| HITS | Hygiene, Integrity, Traceability, and Sharing |

NCD    Non-communicable disease
NaCl    Sodium chloride

## 2.1  Introduction to Advertising and Marketing Strategies in the Food Sector

With reference to the sector of food and beverage (F&B) preparations, three nutritional categories and ingredients—sugars, salt, and milkfat—have been and still are used for enhancing taste, palatability, and satiety, with a few exceptions only when speaking of final products.

However, current marketing and advertising strategies are oriented towards the representation of foods containing low-calorie sweeteners, low-sodium salts, and unrefined vegetable oils. The main reason is that these typologies are represented as unhealthy food ingredients with a strong correlation between excessive intake and obesity, diabetes, inflammation, hypertension, and heart ailments. With reference to this book, an attempt has been made with the aim of underlining chemical properties when correlated to technological food purposes (Chap. 1). The matter of advertising-related aspects is discussed in this chapter, on a general level (Sect. 2.1), and also in detail for each of the three above-mentioned food categories: sugars (Sect. 2.2), salt (Sect. 2.3), and milkfat (Sect. 2.4).

As mentioned in Chap. 1, one of the main challenges for food business operators (FBO) is to solve the matching problem between offered product(s) and/ or service(s) on the one side and current desires of food consumers on the other hand. Food producers, intermediate manufacturers, catering services, and all possible players involved in the F&B supply chain at each (local, regional, national, or global) level have to solve this multifaceted challenge taking also into account regulatory norms and standard certifications where existing (and/or commercially 'obliged' requirements in a determined market, in spite of their voluntary nature).

Each new product or variation/development/enhancement of an existing F&B product has subsequently to be placed on the market and adequately offered to food consumers (at the final stage of the F&B supply chain). This step is critical because it introduces the problem of marketing strategies. Basically, these approaches to the desires and needs of food consumers should be oriented to and/or should favour (Ackroff and Sclafani 1999; Andreyeva et al. 2010; Baker et al. 2020; Chung and Myers 1999; Dias et al. 2020; Dolan and Humphrey 2000; Drewnowski and Monsivais 2020; Gates et al. 2017; González et al. 2022; Ha et al. 2018; Hodgkins et al. 2012; Klosse et al. 2004; Just et al. 2014; Lambert et al. 2024; Mejean al. 2013; Poulton et al. 2010; Sanjari 2017; Xie and Li 2012; Young 2021):

(a) The increase of produced items (in terms of amounts of sold units) with consequent decrease of production costs.
(b) The easy 'translatability' of technical and non-technical information on food labels in a practical and comprehensible marketing/advertising way.

(c) And finally, the continuous improvement of F&B products and services from time to time.

Each of the above-mentioned problems (the list can surely contain other factors) has some critical aspects. First of all, each advertising and/or marketing strategy is based on the enhancement of one or more specific F&B features of a determined product or service if compared with other similar products or services. In other terms, enhanced feature(s) should be correctly advertised, claimed, and previously acceptable (Beverland et al. 2008; Caswell and Mojduszka 1996; Golan et al. 2021; Grunert 2002; Kozup et al. 2003). These attributions are summarisable into a single adjective: 'reliable'. This concept has to be considered with notable attention: each relationship in commercial ambits is based on the resistance of an ethical agreement of reliability and mutual respect between the seller and the buyer. Should this agreement (tacit or express contents are possible) be questioned, the reliability of the whole relationship would be seriously and sometimes irreversibly compromised. The opposite concept of commercial reliability is the well-known '*caveat emptor*' motto ('let the buyer beware', from Latin language) (Calabrese 2017; Jaffe 2003; Pedersen and Neergaard 2006; Steinberg and Luce 2005). For these reasons, each advertising and/or marketing strategy in the F&B sector (all possible FBO are involved, no exceptions) has to face (Antle 1999; Fortin 2022; Fulponi 2006; Haddad and Parisi 2020; Kapelko et al. 2015; Parisi 2022; Pellerito et al. 2019; Puram and Gurumurthy 2023; Rombach and Bitsch 2015; Russo et al. 2023; Segerson 1999; Shukla et al. 2014; Stumpf 2020; Woods et al. 2006):

(1) Regulatory norms with important or sometimes reduced differences, nation by nation (multinational economic areas can impose standardised codes of conduct, technical guidelines, different manufacturing or labelling guidelines, and so on). The ambits of regulation concern four basic pillars of the F&B sector according to the 'Hygiene, Integrity, Traceability, and Sharing' (HITS) acronym (Haddad and Parisi 2020) (i) food safety and public hygiene (including not only microbiological, chemical, and physical dangers, but also undeclared presence of allergens, gluten, genetically modified organisms, and sabotages); (ii) authenticity as opposed to economically motivated adulteration of food frauds, in other terms 'integrity'; (iii) traceability movements on the local, regional, national, or international level; and (iv) food sharing (food supplying by charitable food sharing organisations), where workable

(2) Quality standardised certifications where existing (and/or commercially 'obliged' in a determined market in spite of their voluntary nature), from ISO 9000 norms to F&B-related certifications as recognised by the Global Food Safety Initiative (GFSI) organisation.[1]

Basically, each of these mandatory or voluntary systems is in defence of the F&B consumer. For these reasons, advertising/marketing strategies have to preliminarily consider related implications before proceeding, and the same thing is true for food technologists.

---

[1] Available information can be found at the following web site: https://mygfsi.com/.

After this premise, the following points should still be discussed before considering problems related to salt, sugars, and milkfat. First of all, the increase of sold units with consequent decrease of production costs is directly linked to technical features and commercial factors at the same time. Large-scale distribution is one of the main and basic pillars for commercial success, and this aspect is often linked to another critical problem: the ready availability of raw materials. In other terms, should certain or essential raw materials (including sugars, salt, and/or milkfat) be obtained with strong difficulty in selected commercial areas, the same difficulty would easily be considered when speaking of general distribution of finished F&B products. As a result, F&B preference depends also on non-technical factors (Bourquard et al. 2022; Comba et al. 2013; Engelund et al. 2009; Van Donk 2001).

Secondly, the conversion of technical and non-technical information on food labels in a practical and comprehensible marketing/advertising way for F&B consumers is not a simple work (Andrews et al. 1998; Cantrell et al. 2020; Kozup et al. 2003; Spiteri Cornish and Moraes 2015). Because of different education levels on a local/regional/national scale at least, each successful strategy should emphasise only basic aspects of a certain product, with remarkable preference for easily transferable messages (enhanced safety, freshness, sweetness, crunchiness, ethical meaning, community-related affinity, cultural heritage, etc.). In addition, the conversion of chemical information on a simple level could possibly mislead the consumer, with important effects on the comprehension of the final message. Once more, certain problems should be avoided with a careful design (promotion of useful factors, elimination of difficult concepts). It has to be noted that the following list of names and words—allergens, genetically modified organisms, gluten, ethical messages, religion-related claims, sustainability advertising, organic or fair-trade terms, etc.—are also covered by mandatory norms and/or specific voluntary certification standards.

The continuous improvement of F&B products and services from time to time depends on the success on the market (with already discussed factors) and also on the attention of the 'normal consumer', identified with the average section of a large population sample able to buy commercial goods (Parisi 2012–2022). In this ambit, the problem of 'unhealthy foods' has to be discussed with reference to sugars, salt, and milkfat.

## 2.2   Unhealthy Foods. The Role of Advertising and Marketing Strategies

From the practical viewpoint of food producers, the research for tasteful, sweet, palatable foods (and also for products able to promote satiety) takes into account the role of three basic ingredients or classes of ingredients for food production: sugars, salt, and milkfat (Belc et al. 2019; Delahunty and Havermans 2010; Gibson and Ashwell 2011; Lillford 2016; Low et al. 2014; Mahato et al. 2020; Mukhopadhyay et al. 2020; Yeomans 1998).

However, the use of similar food categories is always correlated with unhealthy food consumption and consequent related overweight and obesity risks, especially when speaking of infant and children. In detail, the association between unhealthy food consumption and the following sub-categories of nutrients has to be taken seriously into account (Cediel et al. 2021; Hartmann et al. 2013; Huffman et al. 2014; Mills et al. 2011; Pries et al. 2019; Rousham et al. 2022):

(1) Added sugars
(2) High salt amount
(3) High intake of saturated and/or *trans*-fats.

Basically, this list cannot be exhaustive. However, nutritionists and other professionals in this ambit seriously take into account the occurrence of unhealthy food consumption in strict connection with the increasing level of sugars, salt, and milkfat. As a result, it appears that the occurrence of a large number of diabetes, hypertension, cardiovascular disorders, and non-communicable diseases (NCD) worldwide may be defined as a pandemic, especially in the developing Nations (Ferreira 2017; Mukhopadhyay et al. 2020; Oakes 2005). As mentioned in Chap. 1, sugars have different functions, including some preservative roles (against microbial spreading). However, their main functions are in the hedonic ambit (enhanced sweetness and attractivity for consumers, etc.) and in the technological ambit with concern to highly desirable features in F&B preparations. These functions are implicitly mentioned on labels, containers, and also in advertising messages, while their unhealthy influence (increased diabesity and inflammation risks, augmented and out-of-scale energy intake, insulin resistance, etc.) are not mentioned. As a result, many F&B preparations with added sugars are familiarly named 'junk foods' (Mukhopadhyay et al. 2020).

With reference to salt, and with the exclusion of its multifaceted food functions, its roles in renal hyperfiltration, hypertension disorders, renal hyperfiltration, kidney diseases, eye disorders, etc., are not explicitly mentioned. Actually, several low-salt campaigns are searching for a partial solution to the problem of excessive salt intake, but good results are not still available on a large scale.

With relation to milkfat, their saturated content and associated NCD are nowadays well known, also because of the related increase of animal cholesterol. However, the matter seems not only include fast or junk foods, but also 'normal' foods. As an example, some researcher might theoretically question the consumption of dairy foods as a food-related problem because of notable amounts of milkfat. On the other side, milkfat has interesting physiological functions, and these roles should not be ignored or underestimated (Escajedo et al. 2010). In this historical context, policy strategies such as the European Union (EU) 'Sugar Tax' have to be considered.

Consequently, it could appear that a correct balancement between food safety recommendations and guidelines on the one side should be given with advertising messages on the other side. On the regulatory level, the recent EU Regulation No 1924/2006 has been dedicated to the harmonisation of allowed nutrition claims and related conditions (Collins and Verhagen 2022). This regulation is briefly considered in the next sections with exclusive reference to sugars, salt, and milkfat-related compounds. However, it should be noted that the situation is still in evolution and probably needs additional improvements, according to recent news (European Parliament and Council 2006; Hieke et al. 2016; Mei 2024; Patterson et al. 2012; Standing Committee on the Food Chain and Animal Health 2007). Anyway, there is a substantial difference between 'nutrition claims', 'health claims', and 'reduction of disease risk claims'. According to the above-mentioned Regulation (Standing Committee on the Food Chain and Animal Health 2007):

(a) 'Nutrition claims' state, suggest, or imply that a F&B product has peculiar and beneficial nutritional properties because of augment, decrease, or absence of energy, or augment, decrease, presence, or absence of particular nutrients
(b) 'Health claims' state, suggest, or imply the existence of a relationship between health and a F&B category, or one specific F&B product, or one of its components
(c) 'Reduction of disease risk claims' state, suggest, or imply a peculiar risk factor in the development of a human disease can be significantly reduced by means of the consumption of a F&B category, or one specific F&B product, or one of its components.

As a consequence, sugars, salt, or milkfat can be considered only when taking into account the above-mentioned differences between allowed claims. The European Regulation is taken as a study example, but other guidelines and related (different) values are available worldwide.

## 2.3  Food-Related Claims and Sugar Content

With reference to the European Union (EU) legislation, the following claims can be considered (European Parliament and Council 2006):

(a) 'Low sugars' (this claim can be used provided that the amount of sugars in the finished F&B product is $\leq 5$ g per 100 g or $\leq 2.5$ g per 100 ml)
(b) 'Sugar-free' (this claim can be allowed on condition that the amount of sugars in the finished F&B product is $\leq 0.5$ g per 100 g or 100 ml)
(c) 'With no added sugars' (this claim is extremely specific, meaning that the F&B product does not contain added monosaccharides, disaccharides, or other ingredient/additive with sweetening functions)
(d) 'Contains naturally occurring sugars' (this claim implies the natural presence of sugars in the F&B product).
(e) 'Reduced (sugars)' (this claim explicitly states that the energy intake for the whole F&B product is reduced if compared with a similar product, and energy reduction is $\geq 30\%$).

A collateral group of energy-related claims concern sweeteners:

(1) 'Low energy'. This claim can be used provided that in a low-energy claimed food the amount of energy is $\leq 40$ kcal (170 kJ) per 100 g (in solid F&B products)

or $\leq$ 20 kcal (80 kJ) per 100 ml (in liquid foods). Should table-top sweeteners be these foods, a maximum value of 4 kcal (17 kJ) per portion with equivalent sweetening power to 6 g of sucrose would be considered

(2) 'Energy free'. This claim can be used provided that the amount of energy is $\leq$ 4 kcal (17 kJ) per 100 ml (in liquid foods) in an energy-claimed food. Should table-top sweeteners be these foods, a maximum value of 0.4 kcal (17 kJ) per portion with equivalent sweetening power to 6 g of sucrose would be considered.

Finally, there are some nutrition claims which could be correlated with sugars (Anonymous 2024):

(a) 'Light' or 'lite' adjectives state that a F&B product can be considered as a product with 'reduced' claims (similarly to 'reduced sugars'). However, an indication of features making this product 'light' or 'lite' has to be proposed (in this ambit, advertising and marketing strategies are critical aspects)

(b) 'Natural or naturally' words mean that a F&B product is naturally able to show peculiar features which can justify a nutrition claim.

## 2.4  Food-Related Claims and Salt Content

With reference to the EU legislation, the following claims can be considered when speaking of salt (European Parliament and Council 2006):

(a) 'Low sodium/salt'. This claim can be used provided that the amount of sodium chloride (NaCl) in the finished F&B product is $\leq$ 0.12 g of sodium, or equivalent NaCl, per 100 g or 100 ml). With reference to the EU, a specific indication concerning natural mineral waters ruled by Directive 80/777/EEC has to be mentioned: in this situation, the maximum allowed amount is 2 mg of sodium per 100 ml

(b) 'Very low sodium/salt'. This claim can be used provided that the amount of salt in the finished F&B product is $\leq$ 0.04 g of sodium, or equivalent NaCl, per 100 g or 100 ml. With reference to the EU, natural mineral waters and other waters are not covered in this way, and such a claim is not allowed

(c) 'Sodium-free' or 'salt-free'. This claim can be used provided that the amount of salt in the finished F&B product is $\leq$ 0.005 g of sodium, or equivalent NaCl, per 100 g or 100 ml

(d) 'No added sodium/salt'. This claim means that (1) the F&B product has been realised without sodium or salt addition, including ingredients containing added sodium or salt, and (2) the amount of salt in the finished F&B product is $\leq$ 0.12 g of sodium, or equivalent NaCl, per 100 g or 100 ml.

Two additional food-related claims may concern sodium/salt:

(1) 'Reduced sodium/salt'. In this ambit, the claimed reduction has to be $\geq$ 25% in comparison with a similar F&B product

(2) 'Natural or naturally' words mean that a F&B product is naturally able to show peculiar features which can justify a nutrition claim.

## 2.5   Food-Related Claims and Fat Content

With reference to the EU legislation, the following claims can be considered when speaking of fat, taking into account that similar requirements concern many fats, including also vegetable fats(Anonymous 2024; European Parliament and Council 2006):

(1) 'Low fat'. This claim can be used provided that the amount of fat is $\leq$ 3 g per 100 g (solid F&B products), or 1.5/ 1.8 g per 100 ml (liquid F&B products and semi-skilled milk, respectively)
(2) 'Fat-free' (this claim can be allowed on condition that the amount of fat in the finished F&B product is $\leq$ 0.5 g per 100 g or 100 ml. Claims as 'X % fat-free' are not allowed in the EU)
(3) 'Low saturated fat'. This claim is extremely specific, meaning that the sum of saturated fatty acids and *trans*-fatty acids in the considered F&B product is $\leq$ 1.5 g per 100 g (solid F&B) or 0.75 g per 100 ml (liquid F&B). In addition, such an amount cannot provide more than 10%-increase of energy intake
(4) 'Saturated fat-free'. This claim states that the F&B product does not contain saturated fat. Analytically, the sum of saturated fat and *trans*-fatty acids is $\leq$ 0.1 g per 100 g or ml
(5) 'Reduced saturated fat' (in this situation, the sum of saturated fatty acids and *trans*-fatty acids in the F&B product has to be $\geq$ 30% if compared with a similar product. In addition, the amount of *trans*-fatty acids has to be equal or less if compared with a similar product
(6) 'High monounsaturated/polyunsaturated fat'. In this situation, a F&B products contain 45% or more of fat derived from monounsaturated/ polyunsaturated fat, and this portion is responsible for more of 20% of energy
(7) 'High unsaturated fat'. This claim means that a F&B products contain 70% or more of fat derived from unsaturated/ polyunsaturated fat, and this portion is responsible for more of 20% of energy.

Finally, there are some nutrition claims which could be correlated with fats:

(1) 'Light' or 'lite' adjectives state that a F&B product can be considered as a product with 'reduced' claims (similarly to 'reduced saturated fat'). However, an indication of features making this product 'light' or 'lite' has to be proposed (in this ambit, advertising and marketing strategies are critical aspects)
(2) 'Natural or naturally' words mean that a F&B product is naturally able to show peculiar features which can justify a nutrition claim.

Basically, each advertising and marketing strategy has to be based on one or more of these labelled messages, when speaking of nutrition claims. On the other side, connection between health and nutritional labels/claims is a difficult matter. More research is needed in this ambit.

# References

Ackroff K, Sclafani A (1999) Palatability and foraging cost interact to control caloric intake. J Exp Psychol Anim Behav Proc 25(1):28–36. https://doi.org/10.1037/0097-7403.25.1.28

Andrews JC, Netemeyer RG, Burton S (1998) Consumer generalization of nutrient content claims in advertising. J Mark 62(4):62–75. https://doi.org/10.1177/002224299806200405

Andreyeva T, Long MW, Brownell KD (2010) The impact of food prices on consumption: a systematic review of research on the price elasticity of demand for food. Am J Pub Health 100(2):216–222

Anonymous (2024) Nutrition claims. Directorate-General for Health and Food Safety, Brussels. Available https://food.ec.europa.eu/safety/labelling-and-nutrition/nutrition-and-health-claims/nutrition-claims_en. Accessed 29th May 2024

Antle JM (1999) Benefits and costs of food safety regulation. Food Pol 24(6):605–623. https://doi.org/10.1016/S0306-9192(99)00068-8

Baker P, Machado P, Santos T, Sievert K, Backholer K, Hadjikakou M, Russell C, Huse O, Bell C, Scrinis G, Worsley A, Friel S, Lawrence M (2020) Ultra-processed foods and the nutrition transition: global, regional and national trends, food systems trans-formations and political economy drivers. Obes Rev 21(12):e13126. https://doi.org/10.1111/obr.13126

Belc N, Smeu I, Macri A, Vallauri D, Flynn K (2019) Reformulating foods to meet current scientific knowledge about salt, sugar and fats. Trends Food Sci Technol 84:25–28. https://doi.org/10.1016/j.tifs.2018.11.002

Beverland MB, Lindgreen A, Vink MW (2008) Projecting authenticity through advertising: consumer judgments of advertisers' claims. J Adv 37(1):5–15. https://doi.org/10.2753/JOA0091-3367370101

Bourquard BA, Berenguer G, Gray AW, Preckel PV (2022) Raw material variability in food manufacturing: a data-driven snack food industry case. Prod Manuf Res 10(1):294–320. https://doi.org/10.1080/21693277.2022.2083030

Calabrese A (2017) Caveat emptor! The rhetoric of choice in food politics. Commun+ 1 6, 1: Article 2. https://doi.org/10.7275/R5CZ35CR

Cantrell K, Li N, Meyers C, Akers C (2020) Misleading or informing? Examining the effects of labeling design on consumers' perception of gluten-free products and wheat safety. J Appl Comm 104(1):2. https://doi.org/10.4148/1051-0834.2309

Caswell JA, Mojduszka EM (1996) Using informational labeling to influence the market for quality in food products. Am J Agric Econ 78(5):1248–1253. https://doi.org/10.2307/1243501

Cediel G, Reyes M, Corvalán C, Levy RB, Uauy R, Monteiro CA (2021) Ultra-processed foods drive to unhealthy diets: evidence from Chile. Pub Health Nutr 24(7):1698–1707. https://doi.org/10.1017/S1368980019004737

Chung C, Myers SL Jr (1999) Do the poor pay more for food? An analysis of grocery store availability and food price disparities. J Consum Aff 33(2):276–296. https://doi.org/10.1111/j.1745-6606.1999.tb00071.x

Collins N, Verhagen H (2022) Nutrition and health claims in the European Union in 2022. Regulatory Focus. Available https://www.raps.org/news-and-articles/news-articles/2022/9/nutrition-and-health-claims-in-the-european-union#citation. Accessed 29th May 2024

Comba L, Dabbene F, Gay P, Tortia C (2013) Open problems in traceability: from raw materials to finished food products. J Agric Eng 44:s2. https://doi.org/10.4081/jae.2013.272

Delahunty CM, Havermans RC (2010) The sensory systems and food palatability. In: Roche HM, Macdonald IA, Schols AMWJ, Lamham-New SA (eds) Nutrition and metabolism, 3rd edn. John Wiley and Sons Ltd, Hoboken and Chichester

Dias ASS, Navas HVG, da Costa AJFP (2020) Design of a continuous improvement model in a Portuguese food industry company–a case study. In: Proceedings of the international congress on engineering—engineering for evolution, KnE engineering, pp 195–208. https://doi.org/10.18502/keg.v5i6.7034

Dolan C, Humphrey J (2000) Governance and trade in fresh vegetables: the impact of UK super-markets on the African horticulture industry. J Develop Stud 37(2):147–176. https://doi.org/10.1080/713600072

Drewnowski A, Monsivais P (2020) Taste, cost, convenience, and food choices. In: Marriott BP, Birt DF, Stallings VA, Yates AA (eds) Present knowledge in nutrition, Volume 2: Clinical and applied topics in nutrition, 11th edn. Academic Press, London, San Diego, Cambridge, and Oxford, pp 185–200. https://doi.org/10.1016/B978-0-12-818460-8.00010-1

Engelund EH, Breum G, Friis A (2009) Optimisation of large-scale food production using lean manufacturing principles. J Foodserv 20(1):4–14. https://doi.org/10.1111/j.1748-0159.2008.00109.x

Escajedo L, Rocandio AM, de Renobales M (2010) Nutrient profiling: are dairy products really unhealthy? In: Romeo Casabona CM, Leire Escajedo San Epifanio L, Emaldi Cirión A (eds) Global food security: ethical and legal challenges. Wageningen Academic, Wagenigemm, pp 228–233. https://doi.org/10.3920/9789086867103_037

European Parliament and Council (2006) Regulation (EC) No 1924/2006 of the European Parliament and of the Council of 20 December 2006 on nutrition and health claims made on foods. Off J Eur Union L404:9–25

Ferreira J (2017) Saturated fat, sodium and sugar in selected food items: a comparison across six European countries. Dissertation, Universidade Nova de Lisboa

Fortin ND (2022) Food regulation: law, science, policy, and practice, 3rd edn. John Wiley & Sons Inc, Hoboken

Fulponi L (2006) Private voluntary standards in the food system: the perspective of major food retailers in OECD countries. Food Pol 31(1):1–13. https://doi.org/10.1016/j.foodpol.2005.06.006

Gates P, Copeland J, Stevenson RJ, Dillon P (2007) The influence of product packaging on young people's palatability rating for RTDs and other alcoholic beverages. Alcohol Alcohol 42(2):138–142. https://doi.org/10.1093/alcalc/agl113

Gibson S, Ashwell M (2011) Dietary patterns among British adults: compatibility with dietary guidelines for salt/sodium, fat, saturated fat and sugars. Pub Health Nutr 14(8):1323–1336. https://doi.org/10.1017/S1368980011000875

Golan E, Kuchler F, Mitchell L, Greene C, Jessup A (2001) Economics of food labeling. J Consum Policy 24(2):117–184. https://doi.org/10.1023/A:1012272504846

González SM, de León CVC, Muñoz YM, Sanchez AP (2022) Continuous improvement for the incursion into the home delivery service of small fast food businesses. Unacienc Rev Estud Investig 15(28):4–13

Grunert KG (2002) Current issues in the understanding of consumer food choice. Trends Food Sci Technol 13(8):275–285. https://doi.org/10.1016/S0924-2244(02)00137-1

Ha OR, Killian H, Bruce JM, Lim SL, Bruce AS (2018) Food advertising literacy training reduces the importance of taste in children's food decision-making: a pilot study. Front Psychol 9:374895. https://doi.org/10.3389/fpsyg.2018.01293

Haddad MA, Parisi S (2020) The next big HITS. New Food Mag 23(2):4

Hartmann C, Siegrist M, Van der Horst K (2013) Snack frequency: associations with healthy and unhealthy food choices. Pub Health Nutr 16(8):1487–1496. https://doi.org/10.1017/S1368980012003771

Hieke S, Kuljanic N, Pravst I, Miklavec K, Kaur A, Brown KA, Egan BM, Pfeifer K, Gracia A, Rayner M (2016) Prevalence of nutrition and health-related claims on pre-packaged foods: a five-country study in Europe. Nutr 8(3):137. https://doi.org/10.3390/nu8030137

Hodgkins C, Barnett J, Wasowicz-Kirylo G, Stysko-Kunkowska M, Gulcan Y, Kustepeli Y, Akgungor S, Chryssochoidis G, Fernández-Celemin L, Storcksdieck genannt Bonsmann S, Gibbs M, Raats M (2012) Understanding how consumers categorise nutritional labels: a consumer derived typology for front-of-pack nutrition labelling. Appet 59(3):806–817. https://doi.org/10.1016/j.appet.2012.08.014

Huffman SL, Piwoz EG, Vosti SA, Dewey KG (2014) Babies, soft drinks and snacks: a concern in low-and middle-income countries? Matern Child Nutr 10(4):562–574. https://doi.org/10.1111/mcn.12126

Jaffe A (2003) Caveat emptor. Am J Med 115(3):241–244. https://doi.org/10.1016/S0002-9343(03)00361-9

Just DR, Sığırcı Ö, Wansink B (2014) Lower buffet prices lead to less taste satisfaction. J Sens Stud 29(5):362–370. https://doi.org/10.1111/joss.12117

Kapelko M, Lansink AO, Stefanou SE (2015) Effect of food regulation on the Spanish food processing industry: a dynamic productivity analysis. PLoS ONE 10(6):e0128217. https://doi.org/10.1371/journal.pone.0128217

Klosse PR, Riga J, Cramwinckel AB, Saris WH (2004) The formulation and evaluation of culinary success factors (CSFs) that determine the palatability of food. Food Serv Technol 4(3):107–115. https://doi.org/10.1111/j.1471-5740.2004.00097.x

Kozup JC, Creyer EH, Burton S (2003) Making healthful food choices: the influence of health claims and nutrition information on consumers' evaluations of packaged food products and restaurant menu items. J Mark 67(2):19–34. https://doi.org/10.1509/jmkg.67.2.19.18608

Lambert EG, O'Keeffe CJ, Ward AO, Anderson TA, Yip Q, Newman PL (2024) Enhancing the palatability of cultivated meat. Trends Biotechnol, in press. https://doi.org/10.1016/j.tibtech.2024.02.014

Lillford PJ (2016) The impact of food structure on taste and digestibility. Food Funct 7(10):4131–4136. https://doi.org/10.1039/C5FO01375E

Low YQ, Lacy K, Keast R (2014) The role of sweet taste in satiation and satiety. Nutr 6(9):3431–3450. https://doi.org/10.3390/nu6093431

Mahato DK, Keast R, Liem DG, Russell CG, Cicerale S, Gamlath S (2020) Sugar reduction in dairy food: an overview with flavoured milk as an example. Foods 9(10):1400. https://doi.org/10.3390/foods9101400

Mei A (2024) Nutrition and health claims regulation, l'eterno incompiuto. Great Italian Food Trade (GIFT), Roma, https://www.greatitalianfoodtrade.it. Available https://www.greatitalianfoodtrade.it/etichette/nutrition-and-health-claims-regulation-leterno-incompiuto/. Accessed 29th May 2024

Mejean C, Macouillard P, Péneau S, Hercberg S, Castetbon K (2013) Consumer acceptability and understanding of front-of-pack nutrition labels. J Human Nutr Diet 26(5):494–503. https://doi.org/10.1111/jhn.12039

Mills JP, Perry CD, Reicks M (2011) Eating frequency is associated with energy intake but not obesity in midlife women. Obes 19(3):552–559. https://doi.org/10.1038/oby.2010.265

Mukhopadhyay S, Goswami S, Mondal SA, Dutta D (2020) Dietary fat, salt, and sugar: a clinical perspective of the social catastrophe. In: Preuss HG, Bagchi D (Eds) Dietary sugar, salt and fat in human health. Academic Press, London, San Diego, Cambridge, and Oxford, pp 67–91. https://doi.org/10.1016/B978-0-12-816918-6.00003-2

Oakes ME (2005) Bad company: the addition of sugar, fat, or salt reduces the perceived vitamin and mineral content of foods. Food Qual Prefer 16(2):111–119. https://doi.org/10.1016/j.foodqual.2004.02.007

Parisi S (2022) The food/packaging synergy: advantages, safety and integrity risks, and possible solutions. Virtual Food Analytical Summit, 04th Oct 2022 – SelectScience, Science House, Church Farm Business Park, Bath BA2 9AP, United Kingdom

Patterson NJ, Sadler MJ, Cooper JM (2012) Consumer understanding of sugars claims on food and drink products. Nutr Bull 37(2):121–130. https://doi.org/10.1111/j.1467-3010.2012.01958.x

Pedersen ER, Neergaard P (2006) Caveat emptor–let the buyer beware! Environmental labelling and the limitations of 'green' consumerism. Bus Strateg Environ 15(1):15–29. https://doi.org/10.1002/bse.434

Pellerito A, Dounz-Weigt R, Micali M (2019) Food sharing—chemical evaluation of durable foods. Springer International Publishing, Cham

Poulton C, Dorward A, Kydd J (2010) The future of small farms: new directions for services, institutions, and intermediation. World Develop 38(10):1413–1428. https://doi.org/10.1016/j. worlddev.2009.06.009

Pries AM, Rehman AM, Filteau S, Sharma N, Upadhyay A, Ferguson EL (2019) Unhealthy snack food and beverage consumption is associated with lower dietary adequacy and length-for-age z-scores among 12–23-month-olds in Kathmandu Valley, Nepal. J Nutr 149(10):1843–1851. https://doi.org/10.1093/jn/nxz140

Puram P, Gurumurthy A (2023) Sharing economy in the food sector: a systematic literature review and future research agenda. J Hosp Tour Manag 56:229–244. https://doi.org/10.1016/j.jhtm. 2023.06.027

Rombach M, Bitsch V (2015) Food movements in Germany: slow food, food sharing, and dumpster diving. Int Food Agribus Manag Rev 18(3):1–24. https://doi.org/10.22004/ag.econ.208398

Rousham EK, Goudet S, Markey O, Griffiths P, Boxer B, Carroll C, Petherick ES, Pradeilles R (2022) Unhealthy food and beverage consumption in children and risk of overweight and obesity: a systematic review and meta-analysis. Adv Nutr 13(5):1669–1696. https://doi.org/10.1093/adv ances/nmac032

Russo C, Sansone M, Colamatteo A, Pagnanelli MA (2023) Sustainability standards: voluntary versus mandatory regulation, JRC130619. Publications Office of the European Union, Luxembourg. https://doi.org/10.2760/598433

Sanjari SS, Jahn S, Boztug Y (2017) Dual-process theory and consumer response to front-of-package nutrition label formats. Nutr Rev 75(11):871–882. https://doi.org/10.1093/nutrit/nux043

Segerson K (1999) Mandatory versus voluntary approaches to food safety. Agribus Int J 15, 1:53–70. https://doi.org/10.1002/(SICI)1520-6297(199924)15:1%3C53::AID-AGR4%3E3.0.CO;2-G

Shukla S, Shankar R, Singh SP (2014) Food safety regulatory model in India. Food Contr 37:401–413. https://doi.org/10.1016/j.foodcont.2013.08.015

Spiteri Cornish L, Moraes C (2015) The impact of consumer confusion on nutrition literacy and subsequent dietary behavior. Psychol Mark 32, 5:558–574. https://doi.org/10.1002/mar.20800

Standing Committee on the Food Chain and Animal Health (2007) Guidance on the implementation of regulation N° 1924/2006 on Nutrition and Health Claims made on food – Conclusions of the Standing Committee on the Food Chain and Animal Health, 14 Dec 2007. Available https://food.ec.europa.eu/system/files/2016-10/labelling_nutrition_claim_reg-2006-124_guidance_en.pdf. Accessed 29th May 2024

Steinberg EP, Luce BR (2005) Evidence based? Caveat emptor! Health Aff 24(1):80–92. https://doi.org/10.1377/hlthaff.24.1.80

Stumpf S (2020) The right to food and private voluntary food standards. Repúb Derecho 5(5):1–22

Van Donk DP (2001) Make to stock or make to order: the decoupling point in the food processing industries. Int J Prod Econ 69(3):297–306. https://doi.org/10.1016/S0925-5273(00)00035-9

Woods M, Thornsbury S, Raper KC, Weldon RN (2006) Regional trade patterns: the impact of voluntary food safety standards. Can J Agric Econ 54(4):531–553. https://doi.org/10.1111/j. 1744-7976.2006.00065.x

Xie X, Li J (2012) Modeling, analysis and continuous improvement of food production systems: a case study at a meat shaving and packaging line. J Food Eng 113(2):344–350. https://doi.org/ 10.1016/j.jfoodeng.2012.05.022

Yeomans MR (1998) Taste, palatability and the control of appetite. Proc Nutr Soc 57:609–615. https://doi.org/10.1079/PNS19980089

Young JS (2021) Measuring palatability as a linear combination of nutrient levels in food items. Food Policy 104:102146. https://doi.org/10.1016/j.foodpol.2021.102146

# Chapter 3
# Case Studies of Human Body Requirements of Sugar, Salt, and Milkfat. A Nutritional and Chemical Review

**Abstract** Sugars, salt, and milkfat have different functions with reference to food and beverage productions, on the technical level. However, the use of sugars, salt, and milkfat products should be evaluated also on the basis of nutritional requirements for human diets. In other words, are these compounds or categories really needed? This chapter is dedicated to three case studies or analyses of the current knowledge concerning sugars, sodium chloride, and milkfat products in food and beverage preparations. With reference to sugars, available recommendations are mainly based on weight-related and disease-related evidences. More research is needed in this research area. With relation to salt, a general trend towards high consumption may be assumed, while lower values than accepted and recommended amounts are not a specific problem so far. Available recommendations are lowering the consumption of salt on a general level, but high consumption can be still observed so far. Finally, available dietary recommendations concerning the necessity of lowering cholesterol intake and chemical profiles of milkfat should be evaluated product by product (by means of the nutritional label) and also considering the class of consumer. More efforts are surely needed in this ambit.

**Keywords** Added sugar · Cholesterol · Dietary intake · Disease · Risk · Salt · Saturated fat

## Abbreviations

DGAC    Dietary Guidelines Advisory Committee
EDA    European Dairy Association
EUFIC    European Food Information Council
F&B    Food and beverage
FDA    Food and Drug Administration
IFST    Institute of Food Science & Technology
IFT    Institute of Food Technologists
MUFA    Mono unsaturated fatty acid

© The Author(s), under exclusive license to Springer Nature Switzerland AG 2024    39
S. Parisi, *Nutrition, Chemistry, and Health Effects of Sugar, Salt, and Milkfat*,
Chemistry of Foods, https://doi.org/10.1007/978-3-031-67395-5_3

NDSR      Nutrition Data System for Research
PUFA      Poly unsaturated fatty acid
SACN      Scientific Advisory Committee on Nutrition
NaCl      Sodium chloride
USDA      United States Department of Agriculture
U.S.      United States of America
WHO       World Health Organization

## 3.1   Introduction to Dietary Lifestyles

As discussed in Chaps. 1 and 2, sugars, salt, and milkfat have different functions with reference to food and beverage (F&B) productions. In detail:

(1) Sugars are linked with claimed sweetening power (and possible possible synergic effects in some products), colour enhancement or modifications (as the result of Maillard reactions), flavour improvement, textural improvement enhancement with associated volumetric augment, good influence on microbial ecology (with associated food durability, acting in contrast with inevitable food alterations according to the First Parisi's Law of Food Degradation), and finally some influence on short term-food intake (Anonymous 2021a; EUFIC 2024; Koivistoinen and Hyvönen 1985; IFST 2022; IFT 2019; Martins et al. 2000; Parisi 2002; Parisi and Luo 2018; Parisi et al. 2019; Singla et al. 2018; Steinert et al. 2011; Vaclavik and Christian 2014a,b)

(2) Sodium chloride (NaCl) is notably considered because of characteristic salty tastes (with possible enhancement of other flavour components in a deter-mined food), F&B processing, manipulation techniques (including sanitisa-tion treatments), and preservation characteristics when speaking of salted foods (Breslin 1996; Huang et al. 2008; Li et al. 2024; Man 2007; Sahni et al. 2023; Thomas-Danguin et al. 2019)

(3) And. last but not least, milkfat products are generally mentioned in F&B products with reference to rheological enhancement, promotion of irreversible or reversible coalescence, ameliorated resistance to lipid oxidation, flavour improvement, textural features (with reference to colour, spreadability, limita-tion of water absorption by protein molecules, etc.), and influence on perceived food satiety in human beings (Abou Samra 2010; Erlanson-Albertsson 2010; Pangborn et al. 1985).

Apart these discourses, the use of sugars, salt, and milkfat products should also be evaluated on the basis of nutritional requirements for human diets. In other words, are these compounds or categories really needed? The next three Sections are dedicated to three case studies or analyses of the current knowledge concerning sugars, NaCl, and milkfat products in F&B preparations.

## 3.2  Is Sugar Needed? What About the 'Right' Amount?

The problem with 'sugars' is that a reasonable dietary intake concerning only this F&B portion, with specific reference to 'added sugars', can be really challenging when speaking of design, proposal, check, and improvement of a theoretical value. In fact, the recent recommendations from the World Health Organization (WHO) have basically encouraged the lowering of added sugars to a maximum 5%-value calculated on the total energy amount from free sugars. However, the currently available food nutrient databases may have different information and data concerning added sugars and their definition. With reference to the United States Department of Agriculture (USDA) and Nutrition Data System for Research (NDSR), different interpretations may cause some confusion when comparing data related to added sugars, and the force of recommendations has to be always evaluated on the basis of the perception of the consumer. As a simple example, the Food and Drug Administration (FDA) has its own definition of sugar, but this definition concerns only mono- and di-saccharides. Consequently, F&B consumers can be confused because of the natural difference between 'added sugars by total sugars' and 'added sugars by available carbohydrates'. Consequently (Clemens et al. 2016; Erickson and Slavin 2015; Mela and Woolner 2018; Palou et al. 2009; Rippe et al. 2017; WHO 2015):

(1) Available studies concerning 'needed' sugars for human diet in the United States of America appear to demonstrate that a 3–8%-energy intake is approximately ascribed to 'added sugars' by total sugars, while free sugars apparently are responsible for 6–16%-energy intake. These data are considered on the basis of USDA sample menu per day

(2) It has been reported that excessively restrictive diets (allowing for a 5%-consumption of 'added sugars' only) could be unsuccessful if not followed by food consumers. At present, such a perspective appears realistic enough, in the United States of America at least.

In conclusion, there are not specific and completely reliable studies concerning our basic question: what about the 'right' amount and intake of sugars for human beings? The only reliable information from the WHO is that a basic recommendation concerning free sugars concerns 10%-energy amount as maximum intake, and it could be lowered to 5% as a conditional recommendation. A 10%-limit is recommended by the Dietary Guidelines Advisory Committee (DGAC) when speaking of added sugars, while the Scientific Advisory Committee on Nutrition (SACN) recommends 5%-energy intake if ascribed to free sugars only. Anyway, these and other recommendations are mainly based on weight-related and disease-related evidences (augment of corpulence data, caries, diabetes, dental problems, increased stroke risks, etc.) (Hood et al. 1977; McGuire 2016; Mela and Woolner 2018; WHO 2015). More research is surely needed in this research area.

## 3.3   Is Salt Needed? What About the 'Right' Amount?

Current health recommendations by the Dietary Guidelines for Americans are mainly based on the concept of adverse outcomes and related risk, similarly to sugars. In the specific ambit and considering sodium intake instead of NaCl, the recommended amount should be between 3 and 5 g of sodium per day, for a normal person (no peculiar advices for specific people groups, communities, and so on). On the other side, a < 3 g-intake could be really low, and a < 5 g-consumption would be also really high: cardiovascular consequences, collateral damages (diabetes), and death are considered risks in this ambit. Basically, sodium is an essential metal for the human body: renal, endocrine, immune, neural, and biochemical systems and processed have to continually maintain sodium levels in blood. Otherwise, cardiovascular risks would easily occur within a limited number of days in these conditions. On the other side, and despite the availability or reliable quantification systems for the definition of sodium intake with a large target group of human subjects, it has been recently reported that 5–10% of the human population worldwide are under 2.3 g of sodium intake per day, while the most part of the considered populations appears to consume 3–6 g of sodium per day. In detail, the average value appears to be 3.95 g per day, while a specific analysis concerning Eastern European Countries, Central and East Asia seems to define a slightly higher value than the average data (4.2 g per day). Anyway, Western Countries such the United States of America appear to be between 4.1 and 4.4 g per day (men) and between 3.0 and 3.5 g per day (women), while the following values: 6.9 g/day (men) and 5.8 g/day (women) are reported in Northern China. As a result, it appears that a general trend towards high consumption may be assumed, while lower values than recommended amounts are not a specific problem so far. Once more, more specific research is needed, also with reference to the analysis of sodium intake levels in specific human categories and the possible reduction of salt levels in processed foods, as recently considered by different mass retailers (Angus 2007; Batcagan-Abueg et al. 2013; Campbell et al. 2015; European Commission 2021; Horikawa and Sone 2017; Loria et al. 2001; Mente et al. 2021; O'Donnell et al. 2020; Strohm et al. 2016).

## 3.4   Is Milkfat Needed? What About the 'Right' Amount?

The problem of milkfat consumption is essentially linked with two aspects:

(a)  The necessity of lowering cholesterol intake (and related cardiovascular effects) because milkfat matter is defined as a hypercholesterolemic source
(b)  The quali-quantitative composition of milkfat.

With reference to the first point, it should be noted that reported hypercholesteromic importance of milkfat is assessed when speaking of saturated fatty acid molecules containing 12, 14, or 16 carbons in the molecular chain. In addition, it has been reported that stearic (C18:0 saturated) acid and oleic (unsaturated) acid have similar effects when speaking of obtaining lower cholesterol levels in plasma. As a consequence, it seems that effective countermeasures in this specific ambit can take into account the modification of fatty acids profiles in the human diet without reducing them (Mohan et al. 2021; Ney 1991).

With concern to the second point, it has to be noted that the consumption for saturated fats and specific milkfat categories strongly depend on specific consumers' categories (elderly, adult, children and adolescents, etc.). As a result, obtained data may be partial and unsatisfactory, even if some broad trend may be observed and certain countermeasures may be suggested. As an example, reported studies for European children and adolescents concerning saturated fat, mono unsaturated fatty acids (MUFA), and poly unsaturated fatty acids (PUFA), have obtained the following reported data in Europe as calculated energy: 10–20% for saturated fatty acids; 10–17% for MUFA; 4–8% for PUFA. On these bases, it may be inferred that Western-style Countries have a specific problem with milkfat consumption, but observed trends may be broad enough and not completely satisfactory (Lambert et al. 2004). The attempt to determine 'good' milkfat amounts with an opposite strategy (comparison between conventional omnivore human subjects and vegan people) may be also unsatisfactory (Larsson and Johansson 2002). Consequently, it has been recently observed that the main part of observational researches aiming at establishing a relationship between milkfat levels from dairy products and health diseases (heart attacks, cardiovascular illness, etc.) have not obtained good results (Huth and Park 2012). Because of the well-known role of milkfat as a deposit of 'emergency energy' in the human body and the importance of physical activity (naturally, a human population includes different types, and a generalisation is not absolutely possible), the following medical advice could be considered (Anonymous 2021b; European Dairy Association 2009): 20.0 g of saturated fat (65–70% in milkfat) on a 2000-cal diet should be assumed as the maximum level for avoiding cardiovascular diseases. Should F&B consumers take this recommendation in practice, they would have several problems because saturated fat is not exactly specific for milkfat only; in addition, the worst part of the matter is that saturated fatty acids are not consumed 'as they are', but included in a notable number of F&B products with or without natural relation to milkfat.

As a final consequence, each dietary recommendation such as FDA's Daily Values for Nutrients (U.S. FDA 2023) or the 'Dietary Reference Values for Food Energy and Nutrients for the United Kingdom' (Neuhäuser-Berthold et al. 1997) should be evaluated product by product (by means of the nutritional label) and also considering the class of consumer (example: child, 1–3 years; female, adult, 19–30 years; etc.) (DGAC 2015). Also, in the case of milkfat, more efforts are surely needed, as demonstrated in many studies and researches along the last 30 years (Castetbon et al. 2009; Garry et al. 1982; Gerrior et al. 1998; Kranz et al. 2007; Jiang et al. 1999; Leslie and Hankey 2015; Rice et al. 2013).

# References

Abou Samra R (2010) Fats and satiety. In: Montmayeur JP, le Coutre J (eds) Fat detection: taste, texture, and post ingestive effects. CRC Press/Taylor & Francis, Boca Raton, FL

Angus F (2007) Dietary salt intake: sources and targets for reduction. In: Kilcast D, Angus F (eds) Reducing salt in foods: practical strategies. Woodhead Publishing Ltd., Cambridge, pp 3–17. https://doi.org/10.1533/9781845693046.1.3

Anonymous (2021a) Parisi's first law of food degradation valuable to establish adequate protocols concerning food durability. Inside Lab Manag 25:1–17 (AOAC International, Rockville, MD)

Anonymous (2021b) Is low-fat or full-fat the better choice for dairy products? Harvard Health Publishing, Harvard. Available https://www.health.harvard.edu/staying-healthy/is-low-fat-or-full-fat-the-better-choice-for-dairy-products#:~:text=Moderation%20is%20key.,saturated%20fat%2C"%20says%20Dr. Accessed 29th May 2024

Batcagan-Abueg AP, Lee JJ, Chan P, Rebello SA, Amarra MSV (2013) Salt intakes and salt reduction initiatives in Southeast Asia: a review. Asia Pac JClin Nutr 22(4):683–697

Breslin PA (1996) Interactions among salty, sour and bitter compounds. Trends Food Sci Technol 7(12):390–399. https://doi.org/10.1016/S0924-2244(96)10039-X

Campbell NR, Correa-Rotter R, Cappuccio FP, Webster J, Lackland DT, Neal B, MacGregor GA (2015) Proposed nomenclature for salt intake and for reductions in dietary salt. J Clinl Hypertens 17(4):247–251. https://doi.org/10.1111/jch.12442

Castetbon K, Vernay M, Malon A, Salanave B, Deschamps V, Roudier C, Oleko A, Szego E, Hercberg S (2009) Dietary intake, physical activity and nutritional status in adults: the French nutrition and health survey (ENNS, 2006–2007). Brit J Nutr 102(5):733–743. https://doi.org/10.1017/S0007114509274745

Clemens RA, Jones JM, Kern M, Lee SY, Mayhew EJ, Slavin JL, Zivanovic S (2016) Functionality of sugars in foods and health. Compr Rev Food Sci Food Saf 15(3):433–470. https://doi.org/10.1111/1541-4337.12194

DGAC (2015) Scientific Report of the 2015 Dietary Guidelines Advisory Committee (DGAC): Advisory Report to the Secretary of Health and Human Services and the Secretary of Agriculture, Appendix E-3.1.A4. Nutritional goals for each age/sex group used in assessing adequacy of USDA Food Patterns at various calorie levels. U.S. Department of Agriculture (USDA), Agricultural Research Service, Washington, D.C. Available https://health.gov/our-work/nutrition-physical-activity/dietary-guidelines/previous-dietary-guidelines/2015/advisory-report/appendix-e-3/appendix-e-31a4. Accessed 29th May 2024

Erickson J, Slavin J (2015) Total, added, and free sugars: are restrictive guidelines science-based or achievable? Nutr 7(4):2866–2878. https://doi.org/10.3390/nu7042866

Erlanson-Albertsson C (2010) Fat-rich food palatability and appetite regulation. In: Montmayeur JP, le Coutre J (eds) Fat detection: taste, texture, and post ingestive effects. CRC Press/Taylor & Francis, Boca Raton, FL

EUFIC (2024) What is the role of sugar in the food industry? European Food Information Council (EUFIC), Brussels. https://www.eufic.org/en/. Available https://www.eufic.org/en/whats-in-food/article/sugars-from-a-food-technology-perspective. Accessed 28th May 2024

European Commission (2021) Defining dietary salt and sodium—examples of implemented policies addressing salt reduction. European Commission, Brussels. Available https://knowledge4policy.ec.europa.eu/health-promotion-knowledge-gateway/defining-dietary-salt-sodium-table-6_en. Accessed 29th May 2024

European Dairy Association (2009) Milk fat. EDA Nutrition Science Fact Sheet. European Dairy Association (EDA), Brussels. Available https://eda.euromilk.org/fileadmin/user_upload/Public_Documents/Nutrition_Factsheets/EDA_Nutrition_Science_Fact_Sheet_-_Milk_Fat.pdf. Accessed 29th May 2024

Garry PJ, Goodwin JS, Hunt WC, Hooper EM, Leonard AG (1982) Nutritional status in a healthy elderly population: dietary and supplemental intakes. Am J Clin Nutr 36(2):319–331. https:// doi.org/10.1093/ajcn/36.2.319

Gerrior S, Putnam J, Bente L (1998) Milk and milk products: their importance in the American diet. Food Rev/Natl Food Rev 21(2):29–37

Hood LF, Wardrip EK, Bollenback GN, Institute of Food Technologists Carbohydrate Division (eds) (1977) Carbohydrates and health. AVI Pub. Co., Westport, CT

Horikawa C, Sone H (2017) Dietary salt intake and diabetes complications in patients with diabetes: an overview. J Gen Fam Med 18(1):16–20. https://doi.org/10.1002/jgf2.10

Huang YR, Hung YC, Hsu SY, Huang YW, Hwang DF (2008) Application of electrolyzed water in the food industry. Food Control 19(4):329–345. https://doi.org/10.1016/j.foodcont.2007.08.012

Huth PJ, Park KM (2012) Influence of dairy product and milk fat consumption on cardiovascular disease risk: a review of the evidence. Adv Nutr 3(3):266–285. https://doi.org/10.3945/an.112. 002030

IFST (2022) Sugars. Institute of Food Science & Technology (IFST), London. Available https:// www.ifst.org/resources/information-statements/sugars. Accessed 28th May 2024

IFT (2019) Sugars: a scientific overview. Institute of Food Technologists (IFT), Chicago. Available https://www.ift.org/career-development/learn-about-food-science/food-facts/food-facts-food-ingredients-and-additives/sugars-a-scientific-overview. Accessed 28th May 2024

Jiang J, Wolk A, Vessby B (1999) Relation between the intake of milk fat and the occurrence of conjugated linoleic acid in human adipose tissue. Am J Clin Nutr 70(1):21–27. https://doi.org/ 10.1093/ajn/70.1.21

Koivistoinen P, Hyvönen L (1985) The use of sugar in foods. Int Dent J 35(3):175–179

Kranz S, Lin PJ, Wagstaff DA (2007) Children's dairy intake in the United States: too little, too fat? J Pediatr 151(6):642–646. https://doi.org/10.1016/j.jpeds.2007.04.067

Lambert J, Agostoni C, Elmadfa I, Hulshof K, Krause E, Livingstone B, Piotr Socha P, Pannemans D, Samartín S (2004) Dietary intake and nutritional status of children and adolescents in Europe. Brit J Nutr 92(S2):S147–S211. https://doi.org/10.1079/BJN20041160

Larsson CL, Johansson GK (2002) Dietary intake and nutritional status of young vegans and omnivores in Sweden. Am J Clin Nutr 76(1):100–106. https://doi.org/10.1093/ajcn/76.1.100

Leslie W, Hankey C (2015) Aging, nutritional status and health. Healthcare 3(3):648–658. https:// doi.org/10.3390/healthcare3030648

Li J, Zhong F, Spence C, Xia Y (2024) Synergistic effect of combining umami substances enhances perceived saltiness. Food Res Int 189:114516. https://doi.org/10.1016/j.foodres.2024.114516

Loria CM, Obarzanek E, Ernst ND (2001) Choose and prepare foods with less salt: dietary advice for all Americans. J Nutr 131(2):536S-551S. https://doi.org/10.1093/jn/131.2.536S

Man CMD (2007) Technological functions of salt in food products. In: Kilcast D, Angus F (eds) (2007) Reducing salt in foods: practical strategies. Woodhead Publishing Ltd, Cambridge, pp 157–173. https://doi.org/10.1533/9781845693046.2.157

Martins SIFS, Jongen WMF, Van Boekel MAJS (2000) A review of Maillard reaction in food and implications to kinetic modelling. Trends Food Sci Technol 11(9–10):364–373. https://doi.org/ 10.1016/S0924-2244(01)00022-X

McGuire S (2016) Scientific report of the 2015 dietary guidelines advisory committee. US Departments of Agriculture and Health and Human Services, Washington, D.C. Adv Nutrition 7, 1:202–204. 10.3945%2Fan.115.011684

Mela DJ, Woolner EM (2018) Perspective: total, added, or free? What kind of sugars should we be talking about? Adv Nutr 9(2):63–69. https://doi.org/10.1093/advances/nmx020

Mente A, O'Donnell M, Yusuf S (2021) Sodium intake and health: what should we recommend based on the current evidence? Nutr 13(9):3232. https://doi.org/10.3390/nu13093232

Mohan MS, O'Callaghan TF, Kelly P, Hogan SA (2021) Milk fat: opportunities, challenges and innovation. Crit Rev Food Sci Nutr 61(14):2411–2443. https://doi.org/10.1080/10408398.2020. 1778631

Neuhäuser-Berthold M, Hauber U, Bruce A (1997) A comparison of dietary reference values for energy of different countries. Z Ernährungswiss 36:394–402. https://doi.org/10.1007/BF0161 7835

Ney DM (1991) Potential for enhancing the nutritional properties of milk fat. J Dairy Sci 74(11):4002–4012. https://doi.org/10.3168/jds.S0022-0302(91)78595-0

O'Donnell M, Mente A, Alderman MH, Brady AJ, Diaz R, Gupta R, López-Jaramillo P, Luft FC, Lüscher TF, Mancia G, Mann JFE, McCarron D, McKee M, Messerli FH, Moore LL, Narula J, Oparil S, Packer M, Prabhakaran D, Schutte A, Sliwa K, Staessen JA, Yancy C, Yusuf S (2020) Salt and cardiovascular disease: insufficient evidence to recommend low sodium intake. Eur Heart J 41(35):3363–3373. https://doi.org/10.1093/eurheartj/ehaa586

Palou A, Bonet ML, Picó C (2009) On the role and fate of sugars in human nutrition and health. Introduction. Obes Rev 10:1–8. https://doi.org/10.1111/j.1467-789X.2008.00560.x

Pangborn RM, Bos KE, Stern JS (1985) Dietary fat intake and taste responses to fat in milk by under-, normal, and overweight women. Appet 6(1):25–40. https://doi.org/10.1016/S0195-666 3(85)80048-9

Parisi S (2002) Profili evolutivi dei contenuti batterici e chimico-fisici in prodotti lattiero-caseari. Ind Aliment 412(41):295–306

Parisi S, Ameen SM, Montalto S, Santangelo A (2019) Maillard reaction in foods. Springer International Publishing, Cham

Parisi S, Luo W (2018) The importance of maillard reaction in processed foods. Springer International Publishing, Cham

Rice BH, Quann EE, Miller GD (2013) Meeting and exceeding dairy recommendations: effects of dairy consumption on nutrient intakes and risk of chronic disease. Nutr Rev 71(4):209–223. https://doi.org/10.1111/nure.12007

Rippe JM, Sievenpiper JL, Lê KA, White JS, Clemens R, Angelopoulos TJ (2017) What is the appropriate upper limit for added sugars consumption? Nutr Rev 75(1):18–36. https://doi.org/10.1093/nutrit/nuw046

Sahni O, Didzbalis J, Munafo JP Jr (2023) Saltiness enhancement through the synergism of pyroglutamyl peptides and organic acids. J Agric Food Chem 72(1):625–633. https://doi.org/10.1021/acs.jafc.3c05911

Singla RK, Dubey AK, Ameen SM, Montalto S, Parisi S (2018) Analytical methods for the assessment of maillard reactions in foods. Springer International Publishing, Cham

Steinert RE, Frey F, Töpfer A, Drewe J, Beglinger C (2011) Effects of carbohydrate sugars and artificial sweeteners on appetite and the secretion of gastrointestinal satiety peptides. Brit J Nutr 105(9):1320–1328. https://doi.org/10.1017/S000711451000512X

Strohm D, Boeing H, Leschik-Bonnet E, Heseker H, Arens-Azevêdo U, Bechthold A, Knorpp L, Kroke A (2016) Salt intake in Germany, health consequences, and resulting recommendations for action. A scientific statement from the German Nutrition Society (DGE). Ernahrungs Umschau 63, 03:62–70. https://doi.org/10.4455/eu.2016.012

Thomas-Danguin T, Guichard E, Salles C (2019) Cross-modal interactions as a strategy to enhance salty taste and to maintain liking of low-salt food: a review. Food Funct 10(9):5269–5281. https://doi.org/10.1039/C8FO02006J

U.S. FDA (2023) Daily value on the nutrition and supplement facts labels—August 2023:1–4. U.S. Food and Drug Administration (FDA), Washington, D.C. Available https://www.fda.gov/food/nutrition-facts-label/daily-value-nutrition-and-supplement-facts-labels. Accessed 29th May 2024

Vaclavik V, Christian E (2014a) Carbohydrates in food. In: Heldman D (ed) Essentials of food science, 4th edn. Springer International Publishing, New York, pp 27–37

Vaclavik V, Christian E (2014b) Sugars, sweeteners. In: Heldman D (ed) Essentials of food science, 4th edn. Springer International Publishing, New York, pp 279–295

WHO (2015) Guideline: sugars intake for adults and children. World Health Organization (WHO), Geneva

# Chapter 4
# Outlook on the Consumption of Sugar, Salt, and Milkfat

**Abstract** The definition of reliable estimations concerning the real consumption of sugars, salt, and milkfat in the current food and beverage market. Basically, the following factors should be included in the analysis: commercial segmentation of food and beverage products; technical differences between edible products; geographical differences; social differences in a well-specified population. Probably, some interesting evaluation could be made on the commercial/economic viewpoint, taking into account the production of processed foods in terms of pure production, distribution, and commercial sales. However, above-mentioned factors can be really challenging in this ambit. On the other side, the problem of social behaviours would be probably diminished in terms of influence on the final result. At present, some economy-related forecast may be supplied with reference to the above-mentioned commodities. With reference to sugars, the production with high prices and related consumption should continue to remain approximately constant or slightly increase in the short period, while more sustainable and health-driven policies are continuously changing the political and also economic context. With concern to salt, current production levels are growing; consequently, food consumption of salt may get low, but food production is constantly growing in general. Finally, with reference to milkfat derivatives, the current trends are extremely influenced by origin. The current situation should not encourage high milkfat consumption in the food sector when speaking of short-term perspectives.

**Keywords** Dairy · Food and beverage market · Forecast · Milkfat · Policy · Salt · Sugar

## Abbreviations

DGAC   Dietary Guidelines Advisory Committee
F&B    Food and beverage
FDA    Food and Drug Administration
SACN   Scientific Advisory Committee on Nutrition

© The Author(s), under exclusive license to Springer Nature Switzerland AG 2024          47
S. Parisi, *Nutrition, Chemistry, and Health Effects of Sugar, Salt, and Milkfat*,
Chemistry of Foods, https://doi.org/10.1007/978-3-031-67395-5_4

NaCl     Sodium chloride
USD      United States Dollars
U.S.     United States
WHO      World Health Organization

## 4.1 The Current Consumption of Sugar, Salt, and Milkfat in the Modern World. A Preliminary Introduction

Chapter 3 has been dedicated to the discussion of possible case studies, researches, and analyses of the current knowledge concerning sugars, sodium chloride (NaCl), and milkfat products in food and beverage (F&B) preparations. Actually, three sligthly different conclusions have been shown:

(1) With relation to sugars, there are not specific and completely reliable results or conclusions concerning the definition of good or acceptable dietary amounts for human beings. The recommended limits—10%- or 5%-energy amount as maximum intake, in accordance with the World Health Organization (WHO), the Dietary Guidelines Advisory Committee (DGAC), the Scientific Advisory Committee on Nutrition (SACN), etc.—do not concern the same target (free sugars, added sugars by total sugars, added sugars by available carbohydrates, etc.). For these reasons, it might be considered (with reference to a very specific area, the United States of America) that 3–8%-energy intake is approximately ascribed to 'added sugars' by total sugars, while 'free sugars' (molecules which are not physically restricted into a cellular envelope, similarly to sugars found in the original vegetable raw materials) are apparently responsible for 6–16%-energy intake. These data cannot be used to calculate the real amount of sugar consumption per person and per day, and the same thing is probably true when speaking of the reliable estimation of sugars inserted into commercial circuits and F&B products. In fact, it has been reported that excessively restrictive diets (allowing for a 5%-consumption of 'added sugars' only) are not realistically possible so far, in the United States of America at least. Finally, it should be remembered that above-mentioned recommendations are mainly based on weight-related and disease-related evidences such as augment of corpulence data, caries, diabetes, dental problems, increased stroke risks, etc. (Hood et al. 1977; McGuire 2016; Mela and Woolner 2018; World Health Organization 2015). More research is surely needed in this contradictory research area (Clemens et al. 2016; Erickson and Slavin 2015; Palou et al. 2009; Rippe et al. 2017)

(2) With reference to salt, available studies and researches have reported that 5–10% of the human population worldwide consume < 2.3 g of sodium per day, while the majority appear to consume 3–6 g of sodium per day. Regional differences should be highlighted (average value: 3.95 g per day; Eastern European

Countries, Central and East Asia, 4.2 g per day; United States of America, 4.1–4.4 g per day and 3.0–3.5 g per day for men and women, respectively), but highest values are also reported to exceed 5.8 g per day. Consequently, a general trend towards high consumption may be assumed, but more specific works are needed, also taking into account specific human categories and the recent trend in the food industry lowering sodium intakes in processed F&B (Angus 2007; Batcagan-Abueg et al. 2013; Campbell et al. 2015; European Commission 2021; Loria et al. 2001; Horikawa and Sone 2017; Mente et al. 2021; O'Donnell et al. 2020; Strohm et al. 2016)

(3)   With concern to milkfat, it has been reported that Western-style Countries have a specific consumption problem, but observed trends may be broad enough and consequently be not completely satisfactory (Lambert et al. 2004). In other words, the majority of observational papers aiming at finding some relationship between milkfat levels from dairy products and health diseases have not obtained reliable results (Huth and Park 2012). The known medical advice which establishes the maximum limit of 20.0 g of saturated fat (65–70% in milkfat) on a 2000-cal diet is mainly based on a risk-based health evaluation. In this ambit, saturated fat is not specifically found in milk and dairy products only (the amount is prevailing in milk, but saturated fatty acids are widely found), but also in processed F&B products with or without natural relation to milkfat (Anonymous 2021; European Dairy Association 2009). Consequently, dietary recommendation such as the United States (U.S.) Food and Drug Administration (FDA)'s Daily Values for Nutrients (U.S. FDA 2023) or the 'Dietary Reference Values for Food Energy and Nutrients for the United Kingdom' (Neuhäuser-Berthold et al. 1997) should be evaluated product by product and also considering the class of involved consumers (DGAC 2015).

Above-mentioned difficulties do not allow us to find and define reliable estimations concerning the real consumption of sugars, salt, and milkfat in the current F&B market. Basically, the following factors are the real problems in this ambit, and the list could be longer than the below-mentioned lines (Barone and Pellerito 2020; Pellerito et al. 2019):

(a)   Commercial segmentation of F&B products
(b)   Technical differences between F&B products, also where similarities are evident and claimed
(c)   Geographical differences, including also dissimilar technical and qualitative standards and certification systems
(d)   Social differences in a well-specified population. After the definition of a specific area, Nation, legislation, market, and so on, the purchasing power of people has to be necessarily taken into account. Otherwise, important data concerning expensive/luxury foods, average-priced F&B products, and cheap products (the category of 'street foods' may be mentioned here, even if there are many reasons of historical and cultural importance) could be compared 'as they are' with the loss of their intrinsic diversity.

Probably, some interesting evaluation could be made on the commercial/economic viewpoint, taking into account the production of processed foods in terms of pure production, distribution, and commercial sales. The association between known production and use data on the one side, and the amount of produced and sold F&B unities could be a possible result. This exercise is basically a traceability test because of the use of different data concerning recognised raw materials (example: raffined monosaccharides or disaccharides, also named with commercial names; table salt or food-grade NaCl; fractionated milkfat derivatives) and the flow of produced and sold F&B unities on the market. Naturally, above-mentioned difficulties—commercial segmentation, technical requirements, geographical identification, different technical and qualitative standards and certification systems, etc.—should be a real challenge in this context. On the other side, the problem of social behaviours (in terms of social differences because of different economic possibilities) would be probably diminished in terms of influence on the final result.

At present, some economy-related forecast may be supplied with reference to the above-mentioned commodities, while the final destination does not appear reliable enough because of discussed difficulties and also the remarkable localisation of main producers of processed F&B products in several Countries. In detail (Abdale and Trimmer 2022; AGRI E 4 2024; Barone 2022; Bolen 2020; Clal.it 2024; Foreign Agricultural Service 2024; Svatoš et al. 2013; Voora et al. 2023):

(1)  With reference to sugars (intended as the whole sector, from the primary production to all possible finished products), the market appears dominated by several Nations only. The economic trends in 2013 seemed extremely drastic when speaking of high prices and consequent low availability worldwide; however, some researchers were confident enough that the whole sugar-related market was and will be linked to economic and politic choices at the same time. In other terms, a short-term prediction could be not reliable enough when speaking of sugars, also because of high prices and the increasing production of sugar commodities at the same time (because the original raw materials can be destined for food and non-food uses, including bioplastics production). The localisation of the main sugar users in South American and South East Asia Countries and in European Nations could be considered as a strong indicator because 60% and more of produced sugar and sugar derivatives appear to be destined to those areas. On the other side, long-term predictions appear unreliable enough so far, with only the forecast that sugar production and prices will continue growing in the next year. However, more recent papers have outlined that the international market of raw cane sugar in the 2010–2022 period has globally fallen from more that 600 United States Dollars (USD) per ton in 2010 to less than 450 USD per ton in 2022. The COVID-19 pandemic has naturally had its own effects, but such a result should demonstrate that economic predictions in such a sector are really challenging, both for producers and final user. In this ambit, there are not reliable facts which could demonstrate that the current use of sugar and sugar derivatives will continue growing in the F&B sector, and sectoral analyses should be considered in the same way. At present, it may be assumed that the

production (with high prices) and F&B utilisation should continue to remain approximately constant or slightly increase in the short period without relation to peculiar F&B sectors, while more sustainable and health-driven policies are continuously changing the political and also economic context

(2)   With concern to salt, the situation could appear confusing enough. Current health-driven policies in many Countries tend to discourage salt consumption: as a consequence, it could be assumed that salt consumption by F&B consumers is falling. However, a specific U.S. study has recently shown that domestic salt production (rock salt, salt in brine, etc.) is slightly grown between 2015 and 2018 at least, and approximately 4% of produced salt is destined to food processing. As a result, it should be assumed F&B use of salt is constant or slightly higher than previous data. With reference to another salt-producing Nation, Italy, it has been recently reported that domestic production of table salt has had several remarkable oscillations between 2014 and 2018, and opposite hypotheses can be made on these bases. The only reliable fact concerns only the clear tendency of National policies, mass retailers, and (naturally) food processors to the lowering of NaCl levels in processed foods. This reflection does not take into account the point that current production levels are growing; consequently, F&B consumption of salt may get low, but F&B production is constantly growing (and F&B consumers seem to increase their expenses with consequent augment of general demand)

(3)   Finally, with reference to milkfat derivatives, the current trends are extremely influenced by origin. In other terms, the U.S. domestic market seems to have growing prices with concern to milk and associated milk/dairy products with the exception of butter (until 2020) because of low milk production growth and notable domestic uses. However, the same report also noted that dairy stocks were declined in 2019 and subsequent 2020 forecasts could be lower than previous data. As a result, F&B uses of milkfat and milkfat derivatives should not be encouraged. Taking into account the notable diversification in this restricted sector, recent EU data concerning milkfat supplies seem to confirm that prices are growing (2014–2024) when speaking of milk delivery, while fat conversion (as butter) appear to decline in the same period (even considering year seasonal variations). In other terms, it appears that milkfat consumption (destination: butter 29%, cream 13%, drinking milk 10%, cheeses 39%, etc.) is not encouraged at present in the F&B sector also for economic reasons. A recent thesis has demonstrated that the use of vegetable oils in mozzarella-imitation cheeses was preferable (if compared with milk-derived raw materials) during 2019 and 2021 on the economic level. As a result, the current situation (and concomitant high F&B prices) should not encourage high milkfat consumption in the F&B sector when speaking of short-term perspectives.

# References

Abdale L, Trimmer LM, III (2022) 2017–2018 Minerals yearbook—Italy [Advance Release]. U.S. geological survey, U.S. Department of the Interior, Reston, VA. Available https://pubs.usgs.gov/myb/vol3/2017-18/myb3-2017-18-italy.pdf. Accessed 30 May 2024

AGRI E 4 (2024) Sugar market situation, 25 April 2024. AGRI E 4—expert group common market organisation on arable crops, European Union, Brussels. Available https://agriculture.ec.europa.eu/document/download/7b38d8b0-e132-40c2-be3f-37657b9402bd_en. Accessed 30 May 2024

Angus F (2007) Dietary salt intake: sources and targets for reduction. In: Kilcast D, Angus F (Eds) (2007) Reducing salt in foods: practical strategies. Woodhead Publishing Ltd. Cambridge, pp 3–17. https://doi.org/10.1533/9781845693046.1.3

Anonymous (2021) Is low-fat or full-fat the better choice for dairy products?. Harvard health publishing, Harvard. Available https://www.health.harvard.edu/staying-healthy/is-low-fat-or-full-fat-the-better-choice-for-dairy-products#:%7E:text%3DModeration%20is%20key.,satura ted%20fat%2C%22%20says%20Dr. Accessed 29 May 2024

Barone M, Pellerito A (2020) Sicilian street foods and chemistry. Springer International Publishing, Cham, The Palermo Case Study

Barone C (2022) Esperienze nel Comparto Agroalimentare di Largo Consumo in Italia. Definizione dei Prezzi su Scala Locale e Globale negli Anni del CoViD-19. Dissertation, Università Telematica Pegaso, Naples

Batcagan-Abueg AP, Lee JJ, Chan P, Rebello SA, Amarra MSV (2013) Salt intakes and salt reduction initiatives in Southeast Asia: a review. Asia Pac J Clin Nutr 22(4):683–697

Bolen WP (2020) Mineral commodity summaries, January 2020—salt, pp. 138–139. U.S. geological survey, U.S. Department of the Interior, Reston, VA. Available https://pubs.usgs.gov/periodicals/mcs2020/mcs2020-salt.pdf. Accessed 30 May 2024

Campbell NR, Correa-Rotter R, Cappuccio FP, Webster J, Lackland DT, Neal B, MacGregor GA (2015) Proposed nomenclature for salt intake and for reductions in dietary salt. J Clinl Hypertens 17(4):247–251. https://doi.org/10.1111/jch.12442

Clal.it (2024) EU: milkfat supply, date: 30 May 2024. Clal.it, Modena. Available https://www.clal.it/en/?section=milkfat_supply_eu. Accessed 30 May 2024

Clemens RA, Jones JM, Kern M, Lee SY, Mayhew EJ, Slavin JL, Zivanovic S (2016) Functionality of sugars in foods and health. Compr Rev Food Sci Food Saf 15(3):433–470. https://doi.org/10.1111/1541-4337.12194

DGAC (2015) Scientific report of the 2015 dietary guidelines advisory committee: advisory report to the secretary of health and human services and the secretary of agriculture, appendix E-3.1.A4. Nutritional goals for each age/sex group used in assessing adequacy of USDA Food Patterns at various calorie levels. U.S. Department of agriculture, agricultural research service, Washington, D.C. Available https://health.gov/our-work/nutrition-physical-activity/dietary-guidelines/previous-dietary-guidelines/2015/advisory-report/appendix-e-3/appendix-e-31a4. Accessed 29 May 2024

Erickson J, Slavin J (2015) Total, added, and free sugars: are restrictive guidelines science-based or achievable? Nutr 7(4):2866–2878. https://doi.org/10.3390/nu7042866

European Commission (2021) Defining dietary salt and sodium—examples of implemented policies addressing salt reduction. European commission, Brussels. Available https://knowledge4policy.ec.europa.eu/health-promotion-knowledge-gateway/defining-dietary-salt-sodium-table-6_en. Accessed 29 May 2024

European Dairy Association (2009) Milk fat. EDA nutrition science fact sheet. European dairy association (EDA), Brussels. Available https://eda.euromilk.org/fileadmin/user_upload/Public_Documents/Nutrition_Factsheets/EDA_Nutrition_Science_Fact_Sheet_-_Milk_Fat.pdf. Accessed 29 May 2024

Foreign Agricultural Service (2024) Sugar: world markets and trade, May 2024:1–9. Foreign agricultural service, United States Department of Agriculture, Washington, D.C.. Available https://apps.fas.usda.gov/psdonline/circulars/sugar.pdf. Accessed 30 May 2024

Hood LF, Wardrip EK, Bollenback GN (eds) (1977) Institute of food technologists carbohydrate division. Carbohydrates and health. AVI Pub. Co., Westport, CT

Horikawa C, Sone H (2017) Dietary salt intake and diabetes complications in patients with diabetes: an overview. J Gen Fam Med 18(1):16–20. https://doi.org/10.1002/jgf2.10

Huth PJ, Park KM (2012) Influence of dairy product and milk fat consumption on cardiovascular disease risk: a review of the evidence. Adv Nutr 3(3):266–285. https://doi.org/10.3945/an.112.002030

Lambert J, Agostoni C, Elmadfa I, Hulshof K, Krause E, Livingstone B, Piotr Socha P, Pannemans D, Samartín S (2004) Dietary intake and nutritional status of children and adolescents in Europe. Brit J Nutr 92(S2):S147–S211. https://doi.org/10.1079/BJN20041160

Loria CM, Obarzanek E, Ernst ND (2001) Choose and prepare foods with less salt: dietary advice for all Americans. J Nutr 131(2):536S-551S. https://doi.org/10.1093/jn/131.2.536S

McGuire S (2016) Scientific report of the 2015 dietary guidelines advisory committee.US Departments of agriculture and health and human services, Washington, D.C. Adv Nutr 7(1):202–204. https://doi.org/10.3945/an.115.011684

Mela DJ, Woolner EM (2018) Perspective: total, added, or free? What kind of sugars should we be talking about? Adv Nutr 9(2):63–69. https://doi.org/10.1093/advances/nmx020

Mente A, O'Donnell M, Yusuf S (2021) Sodium intake and health: what should we recommend based on the current evidence? Nutr 13(9):3232. https://doi.org/10.3390/nu13093232

Neuhäuser-Berthold M, Hauber U, Bruce A (1997) A comparison of dietary reference values for energy of different countries. Z Ernährungswiss 36:394–402. https://doi.org/10.1007/BF01617835

O'Donnell M, Mente A, Alderman MH, Brady AJ, Diaz R, Gupta R, López-Jaramillo P, Luft FC, Lüscher TF, Mancia G, Mann JFE, McCarron D, McKee M, Messerli FH, Moore LL, Narula J, Oparil S, Packer M, Prabhakaran D, Schutte A, Sliwa K, Staessen JA, Yancy C, Yusuf S (2020) Salt and cardiovascular disease: insufficient evidence to recommend low sodium intake. Eur Heart J 41(35):3363–3373. https://doi.org/10.1093/eurheartj/ehaa586

Palou A, Bonet ML, Picó C (2009) On the role and fate of sugars in human nutrition and health. Introduction. Obes Rev 10:1–8. https://doi.org/10.1111/j.1467-789X.2008.00560.x

Pellerito A, Dounz-Weigt R, Micali M (2019) Food sharing—chemical evaluation of durable foods. Springer International Publishing, Cham

Rippe JM, Sievenpiper JL, Lê KA, White JS, Clemens R, Angelopoulos TJ (2017) What is the appropriate upper limit for added sugars consumption? Nutr Rev 75(1):18–36. https://doi.org/10.1093/nutrit/nuw046

Strohm D, Boeing H, Leschik-Bonnet E, Heseker H, Arens-Azevêdo U, Bechthold A, Knorpp L, Kroke A (2016) Salt intake in Germany, health consequences, and resulting recommendations for action. A scientific statement from the German nutrition society (DGE). Ernahrungs Umschau 63(03):62–70. https://doi.org/10.4455/eu.2016.012

Svatoš M, Maitah M, Belova A (2013) World sugar market–basic development trends and tendencies. Agris on-Line Pap Econ Inform 5(2):73–88. https://doi.org/10.22004/ag.econ.152692

U.S. FDA (2023) Daily value on the nutrition and supplement facts labels—August 2023:1–4. U.S. food and drug administration (FDA), Washington, D.C. Available https://www.fda.gov/food/nutrition-facts-label/daily-value-nutrition-and-supplement-facts-labels. Accessed 29 May 2024

Voora V, Bermúdez S, Le H, Larrea C, Luna E (2023) Sugar cane prices and sustainability. Sustainable commodities marketplace series. The International Institute for Sustainable Development, Winnipeg, p 42

World Health Organization (2015) Guideline: sugars intake for adults and children. World Health Organization (WHO), Geneva

# Chapter 5
# Effects of Glucose, Sodium, and Cholesterol Deficiency from a Human Health Perspective

**Abstract** The chemical and nutritional impacts of sugar, salt and milkfat on the food and beverage industry is a challenging topic. The main nutritional categories and ingredients are generally correlated with 'unhealthy' foods when used in excess. On the other side, normal consumers might not too much reduce intake of certain and needed nutrients. This chapter is dedicated to the description of basis reasons for which sugar, sodium, and fat-associated cholesterol should be assumed in a balanced human diet. With reference to sugars, monosaccharides, disaccharides, etc., are needed for a balanced dietary style. The importance of sugars is linked with essential roles when speaking of energy source, storage (as glycogen deposits in the human body), and indirect structural functions. With concern to salt (as expression of sodium), a minimum level is needed with the aim of regulating blood volume at least, and other important metabolic functions rely on sodium availability. Finally, with reference to cholesterol, many metabolic systems and functions are regulated and strictly dependent on cholesterol amount, which is naturally connected with a reasonable amount of milkfat. There are not peculiar reasons to suppose that a good balance between the promotion of public health and sustainable food production cannot be reached.

**Keywords** Cholesterol · Deficiency · Energy · Health · Metabolism · Sodium · Sugar

## Abbreviation

F&B    Food and beverage

© The Author(s), under exclusive license to Springer Nature Switzerland AG 2024    55
S. Parisi, *Nutrition, Chemistry, and Health Effects of Sugar, Salt, and Milkfat*,
Chemistry of Foods, https://doi.org/10.1007/978-3-031-67395-5_5

## 5.1  Introduction to the Problem of Nutritional Deficiences in the Human Diet

The basic aims of this book have been the description of the state-of-art of the food industry and related research with explicit relation to the chemical and nutritional impact of sugar, salt, and milkfat. These main nutritional categories and ingredients are generally correlated with foods containing low-calorie sweeteners, low-sodium salts, and unrefined vegetable oils. These typologies are represented as 'unhealthy food ingredients' on account of a correlation of their excessive intake with obesity, diabetes, inflammation, hypertension, and heart ailments. As a result, derived food and beverage (F&B) products may be associated in this ambit with 'unhealthy' concepts and, consequently, be named 'unhealthy' F&B.

On the other side, normal consumers—impressed of similar advertisements—might not too much reduce intake of certain and needed nutrients. In fact, taking into account the increasing life extension of human population, diseased people might not suffer from deficiency of concerned nutrients like sugar, sodium and cholesterol. This chapter is dedicated to the description of basis reasons for which sugar, sodium, and fat-associated cholesterol should be assumed in a balanced human diet.

## 5.2  Effects of Glucose, Sodium, and Cholesterol Deficiency from a Human Health Perspective

With reference to sugars, there are not reported reasons which could demonstrate that sugar is the only responsible carbohydrate for obesity, overweight, and associated diseases (Anderson 1995). On the contrary, sugars—monosaccharides, disaccharides, etc.—are needed for a balanced dietary style. Consequently, recommendations for a minimum carbohydrate and sugar intake have to be considered not only on the basis of health risk analyses, but also because of the need to maintain certain intake levels for a balanced diet.

As a consequence, why are sugars needed? Basically, sugars have the following critical functions in the human metabolism (Edwards 2016; Murray and Rosenbloom 2018; Sugar Nutrition Resource Centre 2022):

(1)  Energy Source. This function is important when speaking of all body mechanism, including brain functions. Should sugar intake be diminished, brain functions would be inhibited

(2)  Storage. Exceeding carbohydrates can be converted into glycogen which can subsequently be used for prompt energy supply when needed (example: intense sport activities)

(3)  Indirect Structural Functions. Basically, sugars—and also fat matter—are used as energy sources. However, the human metabolism can also rely on amino acids, provided that the amount of available carbohydrates and glycogen are not

sufficient. Should these metabolic pathways be used, the amount of available protein should be progressively decreased with natural decline of muscle masses. As a clear result, dietary intakes ascribed to carbohydrates—including sugars— should be maintained.

With concern to salt (as expression of sodium), it has the following and critical metabolic functions in the human body (Lewis 2023; Wu et al. 2023):

(1) Control of blood volume
(2) Role in functions of kidney and eyes
(3) Potential regulation of energy homeostasis and glucose metabolism (if low-salt intake is considered)
(4) Potential reduction of low-density lipoprotein, triglycerides, and cholesterol amounts (if low-salt intake is considered)
(5) Potential increase of systemic inflammation in heart failure-suffering patients (if low-salt intake is considered).

Some of these functions are negatively affected by high-salt regimes; however, salt is also needed at a minimum level with the aim of regulating blood volume at least. As a clear result, the recommendation of 3–5 g/day amount of salt (Sect. 3.3) is reliable and should be seriously considered.

Finally, with reference to cholesterol, the following points—in favour of its essential functions in the human body—should be considered with attention (Huff et al. 2017; Stevenson and Brown 2009; Tabas 2002; Zampelas and Magriplis 2019):

(1) Regulation of cell membranes integrity and fluidity
(2) Precursor for the production of vitamin D, steroidal hormones (cortisol estradiol, aldosterone, etc.), reproductive-related hormones (progesterone, testosterone, estrogens), and bile acids.

As a final result, many metabolic systems and functions—cellular membranes, stress-related answers, metabolism of calcium, reproductive functions, absorption of fat-soluble vitamins, control of salt and water balance, etc.—are regulated and strictly dependent on cholesterol amount, which is naturally connected with a reasonable amount of milkfat. Once more, a limited but minimum amount of animal fat such as milkfat is surely recommended for a balanced and health diet.

In conclusion, human beings need sugars, salt, and milkfat: the current situation of processed foods is going towards a certain augment of these nutrient profiles, in spite of repeated recommendations and alarms worldwide. However, the above-mentioned situation is not 'fixed' also because of economic and policy-related functions. Consequently, being the basic aim of food nutrition players the necessity of establishing a balanced dietary lifestyle, the best strategy should be to limit sugars-, salt-, and milkfat-related profiles until a minimum recommended intake is reached and effectively workable. Technological and durability-related restrictions have to be considered (Anonymous 2021), but there are not peculiar reasons to suppose that a good balance between the promotion of public health and sustainable food production cannot be reached.

# References

Anderson GH (1995) Sugars, sweetness, and food intake. Am J Clin Nutr 62(1):195S-202S. https://doi.org/10.1093/ajcn/62.1.195S

Anonymous (2021) Parisi'S first law of food degradation valuable to establish adequate protocols concerning food durability. Inside Lab Manag 25:1–17 (AOAC International, Rockville, MD)

Edwards S (2016) Sugar and the brain. Harvard Medica School, Boston, MA. Available https://hms.harvard.edu/news-events/publications-archive/brain/sugar-brain. Accessed 30th May 2024

Huff T, Boyd B, Jialal I (2017) Physiology, cholesterol. StatPearls Publishing, Treasure Island, FL

Lewis JL III (2023) Hypernatremia (high level of sodium in the blood). MDS Manual, Consumer Version. Merck & Co., Inc., Rahway, NJ. Available https://www.msdmanuals.com/home/hormonal-and-metabolic-disorders/electrolyte-balance/hypernatremia-high-level-of-sodium-in-the-blood. Accessed 30th May 2024

Murray B, Rosenbloom C (2018) Fundamentals of glycogen metabolism for coaches and athletes. Nutr Rev 76(4):243–259. https://doi.org/10.1093/nutrit/nuy001

Stevenson J, Brown AJ (2009) How essential is cholesterol? Biochem J 420(2):e1–e4. https://doi.org/10.1042/BJ20090445

Sugar Nutrition Resource Centre (2022) What is the function of sugar in the body? Sugar Nutrition Resource Centre, North Ryde. Available https://www.sugarnutritionresource.org/news-articles/what-is-the-function-of-sugar-in-the-body. Accessed 30th May 2024

Tabas I (2002) Cholesterol in health and disease. J Clin Investig 110(5):583–590. https://doi.org/10.1172/JCI16381

Wu Q, Burley G, Li LC, Lin S, Shi YC (2023) The role of dietary salt in metabolism and energy balance: insights beyond cardiovascular disease. Diabet Obes Metabol 25(5):1147–1161. https://doi.org/10.1111/dom.14980

Zampelas A, Magriplis E (2019) New insights into cholesterol functions: a friend or an enemy? Nutr 11(7):1645. https://doi.org/10.3390/nu11071645